CERN
How We Found the
HIGGS BOSON

CERN
How We Found the
HIGGS BOSON

Michael Krause

 **World Scientific**

NEW JERSEY · LONDON · SINGAPORE · BEIJING · SHANGHAI · HONG KONG · TAIPEI · CHENNAI

Published by

World Scientific Publishing Co. Pte. Ltd.

5 Toh Tuck Link, Singapore 596224

USA office: 27 Warren Street, Suite 401-402, Hackensack, NJ 07601

UK office: 57 Shelton Street, Covent Garden, London WC2H 9HE

Library of Congress Control Number: 2014027852

British Library Cataloguing-in-Publication Data
A catalogue record for this book is available from the British Library.

CERN
How We Found the Higgs Boson

ISBN 978-981-4623-55-1
ISBN 978-981-4623-46-9 (pbk)

Printed in Singapore by Mainland Press Pte Ltd.

Preface

In Search of What Keeps the World Together at Its Heart

> *"The test of all knowledge is experiment."*
> — Richard Feynman, *Lectures on Physics*, 1974

The modern world imposes more and more events, tasks, and information upon us. Everything is fast-paced, and there is hardly any time left to pause and reflect. But reflection and contemplation, our own search for the answers to all the important questions of life, are one of the foundations of human life. Man seeks answers and performs research; he finds things and invents things—curiosity is the most important part of our nature. It is the eternal human search to discover why we are here; why do we exist? Our search for the foundations of the world and the origin of all matter is the subject of this book, which is dedicated to the people at CERN. Why do we want to find out about things—and what brings some people to align their entire lives to this Faustian question: what is it that keeps the world together?

CERN is one of the largest scientific research centers in the world. The most powerful particle accelerator ever built, the Large Hadron Collider (LHC), began operation at the end of 2008. This machine is situated in a huge tunnel 100 meters beneath the surface, crossing the border between Switzerland and France. The LHC provides the highly energetic particles that generate conditions—when colliding head-on in the experiments—similar to those at the beginning of the universe, shortly after the Big Bang. LHC experiments include ATLAS, CMS, ALICE, and LHCb, to name just a few. They are designed to test, to challenge, and maybe to expand our current model of particle physics, the so-called "Standard Model."

This commonly accepted model is the overall picture of our physical world, but it has some fundamental gaps.

In the Standard Model, two questions still remain unanswered:

1) What is the mechanism that gives fundamental particles, the smallest building blocks of our world, their mass?

2) What are the mechanisms and/or the particles responsible for the force of gravity?

The Standard Model offers answers for both of these questions, at least partially. In 1964, the British physicist Peter Higgs calculated that a single particle was responsible for the mass of all particles. The particle, a heavy vector boson, was named the "Higgs boson," and is the last unknown cornerstone of the Standard Model. For gravity, scientists are looking for an as of yet unproven "graviton," a hypothetical particle responsible for gravitation. At CERN, scientists are reasonably certain that their search for the Higgs boson was successful and that the elusive particle really does exist. The force of gravity and its associated graviton are still lingering in as a largely unsecured "extension" to the Standard Model. There exists no concise theory about the phenomenon of gravity on a large, cosmic scale, let alone any practical proof of any particle associated with it.

Besides these two fundamental properties of the world around us, mass and gravity, the common picture of our physical world is missing by far the largest share of the mass/energy content of the universe: dark matter and dark energy. Observations show that they must be there but we don't know what they really are. Dark matter is responsible for the proven fact that rotating systems such as galaxies do not fly apart. It is believed to be something like a "pulp" that cosmic systems are embedded within. In turn, dark energy is then responsible for the proven fact that our universe has been accelerating in its expansion for some billion years now. Dark matter and dark energy together make up the biggest part of the universe, around 96%. According to the current state of physics, the origin and components of both phenomena are still unknown. What the world is really made of—except for maybe love, which is quite important—remains for the most part *terra incognita*. The LHC, the largest and most powerful particle accelerator in the world, was built to explore this unknown terrain. The LHC is conducting research in the highest energy domain, and this huge, complex machine, is opening the doors to a new territory in physics.

The latest technologies, scientific monster machines, and a journey beyond the known physical world are the exciting ingredients of this book. But at the center there is the modern scientist, the initiator and sometimes astonished observer in the quest for the unknown composition of matter. It is not the technology that is important, but the mind that rules it. Not everything is important except what is understandable. These two principles live at the center of this book, which is based mainly on interviews with scientists working at CERN. The goal of this book is to provide a lively portrait of the CERN cosmos—a portrait of the modern scientist pioneering in an innovative era during the early 21st century. This era seems to provide similar conditions to those at the beginning of the last century. Within a rather short period of time, the world around 1900 went through social, political, and scientific quantum leaps. The ingredients for a new era back then were the breakthroughs in atomic physics, political crises, and the rapidly growing social upheaval—alongside the endless curiosity of man on the threshold of a new era, always inspired to seek further discoveries.

The protagonists of this book are the scientists working at CERN. They work in different departments and on different levels, manage projects and experiments, explore previously unknown territories, and think about new theories, yet they all seem to be part of a larger orchestrated process that is targeted at solving an unspecific number of exciting questions. What is it that really keeps the world together at its heart? What does the terrain that the LHC can show us look like? What are the next questions to get to the next answers? The interviews in this book were recorded over a period of several years; they were primarily aimed at the person and the personality of the protagonists to gain insight into the unique world of CERN. Their professional and personal experiences, wishes, thoughts, and insights will bring us closer to that sometimes "curious species of man" (*The New York Times*). If successful, this book will provide a very personal portrait of the necessary human suppositions for fundamental research and new discoveries.

The interviews also give rise to more background information, to content that explains scientifically relevant terms, historic events, and experiments and other fundamental scientific issues. The inserts you will find to some extent consist of quotations. There are basic statements or theoretical laws, which have retained their power and importance as the basis of scientific thinking and research to date. They cover the basic methods of scientific research over the course of the last millennia. It's amazing how the human mind in the course of its development has become increasingly crisp, clear, and unambiguous. Modern scientists tend not to speculate; they're always standing on the shoulders of their predecessors, who have been working on the exploration and scientific explanation of the world around us for centuries.

This book sets out on an exciting journey to explore the basis of CERN and the people who work there in order to give the eternal human quest for the centerpiece of the world a face. All of us are interested in questions like where we came from and where we are going? What is it that we are created from? How does this world really work? Is there eternity and infinity? Or is there an end to it all? What are the right ingredients for this scenario? What is energy? What position do we have in the huge wheelwork of nature, the wheelwork of creation and transience? All of these big human questions are addressed in this book, and perhaps they will be answered, too.

CERN: How We Found the Higgs Boson, documents a historic moment in the history of mankind. The scientists at CERN have been trying to find clear evidence for the existence of the Higgs boson for almost 40 years. It has been christened by Leon Lederman, former director of the American Tevatron, as the "God particle," and the name stuck, mainly because the press liked to call it that. The Higgs boson is the last missing component in the Standard Model in which all of the fundamental particles and forces of nature known to us are put together in one big scheme. In theory, the Higgs boson gives all elementary particles their mass. Whether the Higgs particle found by the CERN experiments really behaves as predicted is so far not really known. It is possible that other mechanisms also apply to this situation or that the Higgs boson is only one representative of a number of previously unknown particles.

> *"Science is human, and humans are never cool. Humans are full of emotion and tragedy."*
>
> — Victor Weisskopf (1908–2002),
> CERN Director-General (1961–1966)

Acknowledgements

I would like to gratefully and sincerely thank:

- Prof. Dr. Rolf-Dieter Heuer, CERN Director-General

- Prof. Dr. Tejinder Virdee, Imperial College London, FRS (Fellow, Royal Society of London), Knight Bachelor

- Dr. Lyndon Rees Evans, LHC project leader, CBE (Commander of the British Empire), FRS

- Dr. Tara Shears, Reader at the University of Liverpool, LHCb experiment

- Prof. Dr. John Ellis, CERN Theory Division, King's College London, CBE, FRS

- Dr. Rolf Landua, Head of Education and Public Outreach Group, CERN

- Prof. Dr. Masaki Hori, ASACUSA experiment, Max Planck Institute of Quantum Optics, Garching

- Prof. Dr. Carlo Rubbia, Nobel laureate in Physics, 1984, Scientific Director, IASS, Potsdam

- Dr. Sebastian White, ATLAS experiment, Zero-Degree-Calorimeter (ZDC) experiment

- Prof. Dr. Albert De Roeck, CMS experiment, University of Antwerp

- Prof. Dr. Jonathan Butterworth, ATLAS experiment, University College London

Contents

1 The History of CERN

"The longer you can look back, the farther you can look forward."
— Winston Spencer Churchill
(1874 – 1965, 1953 Nobel laureate in Literature)

The history of CERN, the European Organization for Nuclear Research, reveals the uniqueness of this scientific endeavor through many interesting details. CERN is the first joint venture in a united Europe, and it might also be considered a symbol of the European ideal. In this book, many of CERN's internal structures will be traced back to their beginnings to make them more understandable. In its initial phase, 60 years ago, the CERN spirit was born. This spirit of scientific excellence can still be felt today.

The Spirit of Europe

Winston Churchill—Member of Parliament since 1900 when he was just 26 and British Prime Minister since 1940—had been defeated in the 1945 British General Election. During the Potsdam Conference, when important decisions about the Allied Nations' (United States, Russia, and Great Britain) next steps regarding Germany and Japan were discussed, Churchill had to hand over power to Labor Party leader Clement Attlee. Churchill continued to be politically very active, and in March 1946 he presented his idea of an "Iron Curtain" in a speech made in the small Missouri town of Fulton (in the USA). This speech ("From Stettin in the Baltic to Trieste in the Adriatic, an iron curtain has descended across the Continent") shaped the politics of the East and West during the Cold War in the 1950s and 60s.

On September 19, 1946, Churchill delivered another famous speech to students at the University of Zurich. In front of the academic youths of neutral Switzerland, Churchill outlined his ideas concerning the future of Europe. Under the rather unpromising title, "The Tragedy of Europe," Churchill presented Europe's course into the future in a very positive and bright light. The visionary sketch was a realistic look into Europe's future and was considered revolutionary at the time. The ex-Chancellor painted the picture of a still powerful, resurgent Europe. At its center, Churchill placed the two most powerful European nations: France and Germany. Churchill insisted on the rapprochement of the two countries, which had been enemies at war with each other until very recently. Churchill also argued for a new, even higher goal, the establishment of "a kind of United States of Europe."

> This noble continent, comprising on the whole the fairest and the most cultivated regions of the earth, enjoying a temperate and equable climate, is the home of all the great parent races of the western world. [...] It is the origin of most of the culture, the arts, philosophy and science both of ancient and modern time. If Europe were once united in the sharing of its common inheritance, there would be no limit to the happiness, to the prosperity and the glory which its three or four hundred million people would enjoy.
>
> It is to re-create the European Family, or as much of it as we can, and to provide it with a structure under which it can dwell in peace, in safety and in freedom. We must build a kind of United States of Europe. [...] If we are to form the United States of Europe, or whatever name it may take, we must begin now.
>
> (EU archives)

Churchill's speech got a lot of international attention; it is often quoted and it is often misunderstood. In essence, the speech deals with the identity and the basis of Europe, which is believed to be united on foundations of justice, freedom, and culture. The speech is not about the creation of a politically united Europe.

At the time when Churchill delivered his visionary speech, much of Europe was still in ruins. Of the three virtues mentioned by the former UK Prime Minister, European culture had suffered the most. But culture in itself is strong and it soon realized a comeback. Churchill's idea was to reunite Europe culturally and to reform the family of European nations. Based on this idea, a joint European project emerged.

CERN—How It All Started

After the end of World War II, European science had lost the worldwide lead and supremacy it had before the war. The whole situation had changed because of the "brain drain" Europe had suffered from the exodus of an entire generation of scientists fleeing the Nazi regime. This event fundamentally changed the scientific community on an international level. After the war, the United States was leading the field, particularly in nuclear and particle physics. The "United States of Europe," as envisioned by Winston Churchill in his speech in Zurich, did not exist at all, and it was, and still is, too difficult to imagine a united Europe.

Despite all these discouraging facts, a small group of European physicists wanted to catch up again with top level international research. Their interests laid in the plan to bring European science, especially particle physics, back to where it had been before WWII. They wanted to reform the European family of nations and to create a European center for nuclear science. This decision, to try and work together to create a common European research institution, was the first step on the long path to the creation of CERN.

CERN Initiatives

CERN exists thanks to the goals set by two groups that joined forces during the reorientation of Europe after 1945: European-minded politicians and particle physicists from all over Europe. The politicians were looking for new ideas that would help facilitate the necessary reconstruction. The physicists—who are very practical by nature when it comes to the realization of their own projects—knew that they had to put all of their national resources together to make European particle physics competitive against the top level research structure in the United States. Only a joint, transnational, and politically sanctioned initiative would be able to raise the large investment capital necessary for a new European nuclear research facility.

Before the war, during a period spanning twenty-odd years—starting with Einstein's *annus mirabilis* in 1905—Werner Heisenberg, Niels Bohr, Erwin Schrödinger, Wolfgang Pauli, and Paul Dirac, among others, laid the foundations of modern nuclear research. The new quantum theories demanded, in addition to the known elementary particles like protons and electrons, a number of hitherto unknown particles. Physicists had been looking for those elusive particles for quite a long time. The decay of highly energetic particles penetrating the Earth's atmosphere and coming from outer space, so-called "cosmic rays," had brought about a new family of particles, called "mesons." This group of mid-weight particles had been predicted by the Japanese physicist Hideki Yukawa (1907–1981, 1949 Nobel laureate in Physics).

However, mesons (from "mesos," the Greek word for "intermediate") are very unstable and decay rapidly. They are created when cosmic rays collide with particles in the higher layers of the Earth's atmosphere. These events unfortunately always happen when the scientist is not there. In order to study cosmic rays (composed primarily of high-energy protons of mysterious origin), and thus facilitate further studies of their atomic structure, the collisions had to be recreated in a scientific laboratory. Basically, European scientists were looking to study the same process that happens in nature through a controlled apparatus. The focus of their research seemed to be quite clear so the name of the first major European physics conference after WWII, the Solvay Conference in 1948, was "Elementary Particles."

After this conference, which was followed by several others, French physicist Louis de Broglie (1892–1987, 1929 Nobel laureate in Physics) submitted the first proposal for a European nuclear research laboratory to be discussed at the European Conference on Culture in Lausanne in December 1949. The Lausanne Conference's main topic was how to promote peaceful cooperation in Europe. Physicists, diplomats, and representatives of scientific institutions, a total of 170 participants from 22 countries, attended the historic conference. They discussed the operating conditions, legal frameworks, financing systems, cooperation possibilities, and addressed other European issues. The Lausanne Conference made a free, transnational spirit of cooperation possible. At Lausanne, a new, modern Europe was created thanks to the neutrality of Switzerland.

During the Lausanne Conference, nuclear physicists met with the political leaders who were essential to getting the project of a European physical laboratory off the ground. Until then, the idea of a pan-European collaboration in physics had no direct or financial support from governments or other authorities. The European Conference on Culture in Lausanne was not only a starting point in the history of a new Europe, it was also the first time after WWII that German diplomats and/ or German physicists were invited to talk with representatives of other European countries about cooperation or anything else. The German delegates certainly realized that the joint European project would be a great chance to polish their rather damaged reputation and that their involvement could facilitate as a kind of reintegration into the international community. The head of the German delegation was Carlo Schmid (1896–1979), one of the fathers of the new, postwar German Constitution. Schmid's speech had a programmatic title: "European is the Creative Spirit."

The initiator of the European Conference on Culture in Lausanne was the Swiss-American writer Denis de Rougemont (1906–1985), who had returned to his home country of Switzerland after several years of residence in the United States. For de Rougemont, Europe was no longer a fantasy. He regarded Europe as a necessity, and in the following years, he campaigned tirelessly for the creation and further development of a new European identity. The Centre Européen de la Culture (CEC) was founded on his initiative in Geneva, in October 1950. This private institution became instrumental in the creation of a pan-European laboratory for nuclear physics, the future CERN. De Rougemont's deep conviction to the European ideal and its cultural values can be felt in each of his own words. A taste of this "spirit" can be felt at CERN even today.

> For what real purposes do we want these cultural means and this education in a common European awareness? From time immemorial Europe has opened itself up to the whole world. Rightly or wrongly, through idealism or ignorance, by virtue of its faith or in pursuing imperialist aims. It has always conceived its civilization as a set of universal values. It is not our purpose to set up a European nation in opposition to the great nations of the east and west, or to desire a synthetic European culture, valid only for ourselves and tuned in upon itself. Our ambition is to contribute to that Union of our countries which will be their sole salvation, through their rebirth of their culture in freedom of the spirit.
>
> — Denis de Rougemont, *Complete Works*, 1994, pp. 95–96

During the Lausanne Conference, de Rougemont pointed out that nuclear physics at the international level were increasingly treated as secret information. In fact, the United States and the United Kingdom were monopolizing nuclear research. After the creation of the atomic bomb and the devastating bombings of Hiroshima and Nagasaki, Europe lagged far behind. At the beginning of the conference, de Rougemont argued for a common European center for nuclear research to keep up with the US and UK institutions in this field. The next item on the Lausanne agenda was Louis de Broglie's proposal. It was read by Raoul Dautry, general manager of the French "Commissariat à L'Energie Atomique" (CEA). De Broglie's proposal basically stated that a collaboration between European states would allow for projects that were impossible to achieve at the national level. During the conference, Dautry would persuade Pierre Auger (1899–1993), Director of the Exact and Natural Sciences Division of UNESCO, of the necessity for a pan-European laboratory for particle physics.

The United Nations Organization for Education, Science, and Culture (UNESCO) was founded on November 16, 1945. The organization was the political stage upon which a multinational European nuclear research center could be created that would be accepted by the US and the UK as well. After the Lausanne conference, the next step came during the fifth UNESCO General Conference in Florence, Italy, in May 1950. The global situation had changed dramatically since the Lausanne Conference; the USSR had successfully detonated its first nuclear weapon in August 1949 and the rather weak European position regarding nuclear research had to be strengthened by the United States.

"I think physicists are the Peter Pans of the human race. They never grow up, and they keep their curiosity."

— Isidor Isaac Rabi (as quoted in "The Atomic Scientist, the Sense of Wonder and the Bomb", by Mark Fiege, *Environmental History*, 12, 2007)

Isidor Isaac Rabi
(Nobel Foundation, 1944)

During the UNESCO Conference in Florence, the American particle physicist Isidor Isaac Rabi (1898–1988, 1944 Nobel laureate in Physics "for his resonance method for recording the magnetic properties of atomic nuclei") smuggled the idea of a new European laboratory into the agenda. Rabi had met with the Italian physicist Edoardo Amaldi (1908–1989) weeks before and liked his ideas for the European laboratory. Rabi had already worked on a similar project, the American particle accelerator "Cosmotron" in Brookhaven, Long Island. The new American nuclear research center near New York City was a joint project of nine major universities (e.g., Columbia, MIT, and Harvard, among others). Rabi had also been a key figure in the Manhattan Project, the successful American effort to build an atomic bomb. As a member of the United States Atomic Energy Commission Rabi had substantial political influence.

At the end of the Florence Conference, Pierre Auger, Edoardo Amaldi, and Isidor Rabi proposed a resolution authorizing UNESCO "to assist and encourage the formation of regional centres [sic] and laboratories in order to increase and make more fruitful the international collaboration of scientists in the search for new knowledge in fields where the effort of any one country in the region is insufficient for the task." The proposal was unanimously adopted by the participants. It provided the political framework for the establishment of a European physics laboratory, a large-scale project with significant impact.

Pierre Auger served as UNESCO director in 1948, and as a former director of the French CEA, he had close ties to the French scientific community. Both Rabi and Amaldi had worked for the US Atomic Energy Commission; they knew each other very well. These were excellent ingredients for a good cooperation. Right after the Florence Conference, Auger began to use his political network to promote the idea of a European physics laboratory. During the following weeks, Amaldi traveled to the United States to inspect the construction of the Cosmotron at Brookhaven, a particle accelerator with a hitherto unattained particle energy of up to 3 gigaelectron volts (GeV). With this powerful machine, it would become

Pierre Auger, Edoardo Amaldi, and Lew Kowarski, Paris, 1952. (© CERN 1952, CERN-HI-5202016)

much easier to investigate phenomena in particle physics and to gain better insight into the internal structure of the atomic core (nucleons). When Amaldi visited the impressive facility in Brookhaven—the accelerator had a diameter of 23 meters—he could only say one word, "Colossal!"

CERN's History

- Starting point: The progress and development of nuclear physics requires a large particle accelerator, causing enormous costs.
- European physicists develop a plan for a common European physics research center.
- The Founding Fathers of CERN: Isidor Rabi, Pierre Auger, Niels Bohr, Edoardo Amaldi, Raoul Dautry, Louis de Broglie, and Lew Kowarski.
- Objective is the better understanding of the structures of atoms and elementary particles.

The Centre Européen de la Culture organized another meeting in Geneva in December 1950. There, Pierre Auger presented a vague building plan for the new laboratory. The layout was set up not to work with nuclear reactors but with particle accelerators. It was similar to the US plant in Brookhaven, the Swedish apparatus in Uppsala, and the British machine at Harwell, which was essentially aimed at producing plutonium, the necessary material in building an atomic bomb.

Great Britain had the necessary know-how, but—officially—showed no interest whatsoever in the European initiative, despite numerous informal contacts. Until then, British physicists had not attended the meetings and conferences of the Auger/UNESCO group. In the UK, the favored plan was to stay with their own facilities. The UNESCO project was regarded rather skeptically and critically. Herbert W.B. Skinner (1900–1960), professor at the University of Liverpool, characterized such intentions as "high-flying and crazy ideas." Skinner favored a plan for the construction of an accelerator in Liverpool. The political climate in the UK was rather negative regarding the European lab. At the same time, there were many high-ranking British scientists who were interested in a European laboratory for particle physics. Sir John Cockcroft, Director of the UK Atomic Energy Research Establishment (AERE at Harwell) even sent his colleague, Frank Goward (1919–1954), to Geneva as a British observer. Goward was later appointed deputy director of the Proton Synchrotron Group at CERN in 1952.

Despite British ambivalence, a resolution was proposed in Geneva for the establishment of a new laboratory and the construction of a particle accelerator. The machine's energy was to be larger than that of the plant in Brookhaven and twice as powerful as the accelerator in Berkeley, USA (the "Bevatron"). Another important step was when representatives of Italy, France, and Belgium donated $10,000 to the project. This seed capital laid the foundation for the yet to be established European institution for particle physics. With this financial capital, Pierre Auger set up an office at UNESCO and in May 1951, a group of experts was chosen for the development of a more detailed plan to be presented at the next UNESCO conference. The Members of the Board of Consultants were Edoardo Amaldi (Italy), Paul Capron (Belgium), Odd Dahl (Norway), Frans Heyn (the Netherlands), Lew Kowarski and Francis Perrin (France), Peter Preiswerk (Switzerland), and Hannes Alfvén (Sweden).

Odd Dahl (1898–1994) had been a pioneer in a number of different scientific fields. He had been a pilot in Roald Amundsen's North Pole expeditions; in the United States, he had worked for the Carnegie Institute. Dahl emphasized another important feature of the new laboratory, and CERN continues to be proud of this function within the European academic world:

> It may also be pointed out that a modern nuclear physics laboratory is, in scope, a universal laboratory, even if the final aim is for highly specialized knowledge. The projected laboratory will thus serve as a training centre [sic] in coordinated research covering physics, biophysics, chemistry, bio-chemistry, technology and medicine in its basic and applied forms, creating a type of research worker adaptable to industrial research in his own home country. In this connexion [sic], it might be a good thing if as a matter of principle the laboratory were staffed, to a large extent, by transient or semi-transient personnel who would afterwards return to institutions in their own countries.
>
> (*UNESCO and Its Program, XI,* 1954)

The Ambivalent Role of the British

In the UK, the development of domestic facilities in collaboration with the Niels Bohr Institute was favored. Many British scientists were in close contact with the

Copenhagen Institute and detailed plans for a British-Danish cooperation were under way. Naturally, the Auger/UNESCO group plan was viewed rather skeptically. British scientists feared that they would lose their technological edge and believed that the idea of a European laboratory was based not on the experience and expertise of the participants but on their courage and enthusiasm for taking risks.

One of the key players on the British side was Sir James Chadwick (1891–1974, 1935 Nobel laureate in Physics "for the discovery of the neutron"). Some years later, Chadwick's discovery had led to successful nuclear fission experiments, which were conducted by German physicist Otto Hahn (1944 Nobel laureate in Chemistry "for his discovery of the fission of heavy nuclei"). During the war, Chadwick—a student of Ernest Rutherford—had worked on the Manhattan Project. Chadwick regarded the Auger/UNESCO group, especially the participation of Germany, with extreme skepticism. In a letter to his colleague, Dr. C. David King, Chadwick outlined the ambivalent British position regarding the European project:

- No direct participation in the project
- No support either in personnel or in financial terms
- No total withdraw from the Auger plan
- Only informal help by word and deed, especially regarding the technical design of the laboratory and design of the machines

Chadwick's view of the situation makes the overall behavior of British nuclear physicists regarding the burgeoning European competitors easier to understand. The envisaged cooperation with the Niels Bohr Institute contributed largely to the British ambivalence concerning the whole situation. The London-Copenhagen axis meant that the two parties were going to go their own way. At the same time, British scientists did not want to miss an opportunity, and they were ready to help their European colleagues. The International "Codex of Science," established after WWI, manifested in a comprehensive professional exchange across borders with informal agreements and general technical and scientific support.

The deeper cause of British opposition was the fact that official involvement in the Auger/UNESCO project would have gone against Britain's national interests. At Harwell, the nuclear research facility in Oxfordshire, a 170 MeV (megaelectron volts) Synchro-Cyclotron was already in operation together with an electron synchrotron and a linear accelerator. At other universities across the country: in Glasgow, Liverpool, and Birmingham, similar plants were either under construction or already operating. The French support for a joint European project was considered an attempt to gain superior technological knowledge.

Reasons for British opposition to the idea of a joint European laboratory for particle physics:

- British nuclear physics was superior to the existing scientific research facilities on the continent. Naturally, the British wanted to stay ahead. If the French wanted to build a research laboratory for nuclear physics, they could continue to do so with the help of other nations.
- Domestic political situation: After WWII, it was preferable to work with existing domestic institutions rather than with new international institutions. The main motivation here was tradition and the fear of foreign, non-British entities.
- The British Labour Party, which was totally opposed to European ideals and politics, had won the 1945 election, ousting Chancellor Winston Churchill. Labour's major goal was to strengthen the UK's traditionally good relations with the US, especially regarding research in nuclear physics.

CERN—An Idea Takes Shape

The next UNESCO conference was held in Paris in July 1951, where the Auger plan was favored over the proposals from the UK and Denmark. A UNESCO studies group (composed of Cornelis Bakker, Odd Dahl, Frank Goward, and others) reviewed the existing plans with the help of some British scientists. Due to the costs, building the most powerful particle accelerator in the world was actually not the first option. According to the new plan, two smaller machines were on the agenda located either in Copenhagen or in Geneva. In addition, the group proposed the establishment of an interim organization to develop the necessary construction and budget plans for the UNESCO project.

Niels Bohr (1885–1962, 1922 Nobel laureate in Physics "for his services in the investigation of the structure of atoms and of the radiation emanating from them") had not yet attended any of the Auger/UNESCO group's meetings. However, on June 9, 1950, shortly after the UNESCO conference in Florence, Bohr made his position quite clear when he wrote an open letter to the UN. Bohr expressed his skepticism and hopes, which were based "entirely on [his] own responsibility and without consultation with the government of any country":

I find it difficult to convey with sufficient vividness the fervent hopes that the progress of science might initiate a new era of harmonious cooperation between nations, and the anxieties lest any opportunity to promote such a development be forfeited.

The ideal of an open world, with common knowledge about social conditions and technical enterprises, including military preparations, in every country, might seem a far remote possibility in the prevailing world situation.

Still, not only will such relationship between nations obviously be required for genuine cooperation on progress of civilization, but even a common declaration of adherence to such a course would create a most favorable background for concerted efforts to promote universal security.

At the end of summer 1951, Pierre Auger visited Niels Bohr in Copenhagen to get the doyen of nuclear physics on board with the European project. During the meeting, Bohr expressed two main concerns about the European initiative. He mentioned the huge financial dimensions of a large accelerator program, on the one hand, and questioned its future viability. On the other hand, he felt that a joint international laboratory should be developed out of an existing institution—Bohr was thinking of his Institute for Theoretical Physics at Copenhagen University, which he had led since 1920. During WWII, an annex to the institute had been built, and this new building could have housed the future European research center.

Auger's interim group envisioned another model for future European cooperation. At a meeting in Paris, in November 1951 a new proposal was discussed. It stated that all research groups should continue to work at their home institutions and universities: Bakker in the Netherlands, Kowarski in France, Dahl in Norway, Bohr in Copenhagen, and Amaldi in Rome. This proposal, in hindsight, would have been totally impractical and CERN would not exist as it does today.

Despite all previous concerns and political shyness, discussion was pushed forward by the British. After Auger's meeting with Bohr, James Chadwick tried to make Bohr's ideas more popular among his colleagues in England—ideas for a laboratory jointly operated by the UK and the Niels Bohr Institute, and based in Copenhagen. The reactions, however, were unenthusiastic. British scientists preferred to work at their own plants (e.g., Harwell and Glasgow) rather than building a new one somewhere abroad. However, Sir George Thomson (1892–1975, 1937 Nobel laureate in Physics), a professor at Imperial College London, showed a strong interest in the plans of the Auger group and was invited to attend the next meeting of the Interim Group as an observer.

Intergovernmental Meeting of UNESCO in Paris, December 1951

The sixth session of the UNESCO General Conference was held in Paris from December 17–21, 1951, and it was perhaps the most important conference in the history of CERN. Representatives from 21 countries attended and although no final decisions were made, considerable progress was made regarding the layout of a new pan-European Laboratory for Particle Physics and to financing commitments.

Bruno Ferretti (1913–2010) was Italy's representative at the Paris Conference. Before the war, Ferretti had worked with Enrico Fermi, who had immigrated to the United States in 1938 to become one of the fathers of the atomic bomb. Ferretti was also a close friend to Edoardo Amaldi. During the conference, Ferretti presented a more detailed plan for the new European nuclear physics laboratory on the basis of a large accelerator for elementary particles. This plan was discussed in depth and controversially. Then Sir George Thomson made his statement and to no one's surprise he stood up for British interests. In his speech, Thomson pointed out that Britain had already invested heavily in this field of research and technology, i.e., in the construction of machines, magnets, and accelerators. Instead of focusing on horribly expensive new facilities, Thomson pointed out that there was already a large new synchrotron under construction that could be used for further cooperative research in Liverpool, England. The French delegate, Francis Perrin, disagreed. He believed the construction of a joint laboratory should not be delayed in the face of the European situation. Otherwise, young European scientists would move to the United States to work with the new, more powerful and thus more attractive accelerators there.

The German representative at the Paris Conference was Werner Heisenberg (1901–1976, 1932 Nobel laureate in Physics, 1932 "for the creation of quantum mechanics"), and his presence was discussed quite controversially. During WWII, Heisenberg had been director of the German Kaiser Wilhelm Institute of Physics and as such he had worked on the Nazi atomic bomb project. Heisenberg's personal relationship with Niels Bohr also had bitter traces from the past. Heisenberg had been Bohr's assistant in the 1920s, so they knew each other pretty well. Nevertheless Heisenberg's ominous visit to Copenhagen during the war, in September 1941, had left the two scientists with two different kinds of remembrances. There are no written records of this historic meeting in which Heisenberg—according to Bohr's memory—had offered Bohr the cooperation of Nazi Germany. Several months later, in 1943, Bohr would flee from Denmark. The British flew him out. From the UK, Bohr went to the United States where he helped the American atomic bomb program to succeed. Much later, in 1958, Bohr made remarks about the meeting with Heisenberg. Until then, Bohr had kept silent about Heisenberg's visit, because "mankind was at issue in which, despite our personal friendship, we had to be regarded as representatives of two sides engaged in mortal combat."

In postwar Germany, Heisenberg had been appointed director of the Max Planck Institute for Physics in Göttingen. At the conference in Paris in 1951, Heisenberg was the official representative for the Federal Republic of Germany. He seemed to be quite skeptical about a common and quite costly European nuclear physics laboratory. Heisenberg pointed out Germany's financial situation and the so far unconvincing planning endeavors:

> Our country is in [an] extremely difficult economic position and I am not entitled at the present to commit our government to any expense in this direction. [...] One should not just try to copy one of the big American machines.

Despite all of the problems, the Paris Conference proved that the majority of the delegates favored the establishment of a new European laboratory. All twelve participating nations agreed to accept the proposal of the Dutch delegation for the formation of an interim organization and a document was signed. This formal act furthered the project a great deal but a catchy name for the new laboratory was still missing. At the time, the initiative operated under the name "Council of Representatives of European States for Planning an International Laboratory and Organizing other Forms of Co-operation in Nuclear Research"; it was a precise description but perhaps a bit too long.

The final document of the Paris Conference transformed an independent scientific initiative into a multinational concept supported by UNESCO and twelve other nations. At the end of the conference, France, Switzerland, Italy, Belgium, and Yugoslavia agreed to support it with $150,000 set aside for further planning. Concrete realization potential for the whole project thus grew immensely and even the international media was hooked. On December 21, 1951, the *New York Herald Tribune* published an article under the headline "Europe Laboratory May Get Five Billion Volt Cyclotron," stating that "Physicists from 12 nations announced their intention last night of creating a European Nuclear Research Laboratory, including a 5,000,000,000-volt cyclotron matching the most powerful machine now known to be under construction at Berkeley, Calif." The most important message to American readers of the newspaper was "that if their governments indorsed [sic] their project and it became a reality, it would have no military value."

UNESCO General Conference, Sixth Session, Paris, December 1951

- Concrete proposal for a large accelerator for elementary particles made by Bruno Ferretti.
- 12 participating States: Belgium, Denmark, France, Great Britain, Greece, Italy, the Netherlands, Norway, Sweden, Switzerland, West Germany, and Yugoslavia.
- Interim organization commissioned by UNESCO receives financial support.
- A preparatory period of 12 to 18 months to further develop accelerator and facility plans.

The Birth of CERN

Two months later, on February 15, 1952, the second session of the UNESCO intergovernmental conference was held in Geneva. The interim agreement was finalized and the organization was renamed (in French) to "Conseil Européen pour la Recherche Nucléaire," or "European Council for Nuclear Research." Thus the ambitious European project for particle physics finally got an official and usable name—and the acronym "CERN" was born.

The Geneva Agreement was immediately adopted by the Federal Republic of Germany, the Netherlands, and Yugoslavia. Eight other countries (i.e., France, Belgium, Italy, Norway, Greece, Sweden, Switzerland, and Denmark) had to wait until the plan was ratified by their respective parliaments over the forthcoming months. The British still considered themselves ahead of the other European countries in terms of particle physics and insisted on their own nuclear research policy. They did not sign the agreement but kept the door open with regular payments of their significant contribution to the CERN budget. Denmark strongly insisted on Copenhagen (the Niels Bohr Institute) as the location for the laboratory while all of the other delegates voted for the other proposed location, i.e., Geneva.

First meeting of the CERN Council in Geneva on February 15, 1952: Sir Ben Lockspeiser, Amaldi, Bloch, Kowarski, Bakker, and Niels Bohr. (© CERN 1952, CERN-HI-5201001)

The Geneva Agreement provided CERN with a development budget of 2,081,945 Swiss francs or 487,000 US dollars. The participating States committed themselves to contribute to the CERN Council: Belgium with 137,045 Swiss francs; Denmark with 60,300 Swiss francs; France with 555,900 Swiss francs, the Federal Republic of Germany with 332,500 Swiss francs; Italy with 213,000 Swiss francs; the Netherlands with 61,100 Swiss francs, Norway with 40,300

In your reply, please refer to :
En répondant, veuillez rappeler :
N°

Genève, 15 février 1952

Professor I. Rabi,
Columbia University,
New York, N/Y.

We have just signed the Agreement which constitutes the official birth of the project you fathered at Florence. Mother and child are doing well, and the Doctors send you their greetings.

Letter from the CERN Council to Isidor Rabi, February 15, 1952. (© CERN 1952)

Swiss francs; Sweden with 98,900 Swiss francs; Switzerland with 138,000 Swiss francs; the United Kingdom with 364,400 Swiss francs; and Yugoslavia with 52,000 Swiss francs.

To express the joyous mood of the day, Pierre Auger, Edoardo Amaldi, Niels Bohr, and more than a dozen enthusiastic nuclear physicists thanked Isidor Rabi for his initial help in creating CERN. Under the UNESCO letterhead, the gentlemen thanked their American *spiritus rector* with intelligent humor.

Cosmic Rays from Outer Space

Werner Heisenberg signed the Geneva Agreement as representative of the Federal Republic of Germany. Through this act, Germany once again began to play a respected role in international science. *Der Spiegel*, a German magazine, featured Heisenberg on its cover page. The magazine wrote that the 51-year-old had great will power and courage that would make him one of the "clearest thinkers of the present generation, in the style of Albert Einstein."

In early summer 1925, during the *Sturm und Drang* (storm and stress) years of quantum physics, Heisenberg had discovered the foundations of quantum mechanics at the young age of 24. Heisenberg's uncertainty principle states that the position (time) and momentum (energy) of any particle cannot be determined with absolute accuracy. In Heisenberg's own words:

> The more precise the measurement of position, the more imprecise the measurement of momentum, and vice versa.
>
> (*Zeitschrift für Physik*, 43, 1927)

In his theory, Heisenberg replaced Bohr's elliptical electron orbits around the atom's core with waves and probabilities. This provided the theoretical basis for the new matrix of modern quantum physics, the foundation of any modern physics research. Heisenberg's work on quantum theory won him the 1932 Nobel Prize in Physics. Albert Einstein, another guru of physics, took a completely different view of Heisenberg's uncertainty principle. Einstein, writing to Max Born on December 4, 1926, said:

> The theory yields a lot, but it hardly brings us any closer to the secret of the Old One. In any case I am convinced that He does not throw dice.

Heisenberg realized that further research into the internal structure of the atom would only be possible with large accelerators to collide the particles at very high energies. He had intensively studied the properties of cosmic rays since 1946 ("Rays from space," *Der Spiegel*, 24/1952) and regarded them as "the ideal testing ground of atomic physicists." This inevitably meant the construction of large accelerator facilities, such as CERN.

The CERN Council held its first official meeting in Paris from May 5–8, 1952. In the meantime, plans had been further developed but the accelerator's location remained unclear. Besides Geneva and Copenhagen, Arnhem in the Netherlands, Liverpool and Paris were mentioned. At the meeting in Paris, key personnel decisions were made:

- Secretary-General: Edoardo Amaldi (Italy)
- Director of Proton Synchrotron (PS) Division: Odd Dahl (Norway)
- Director of Synchro-Cyclotron (SC) Division: Cornelis Jan Bakker (the Netherlands)
- Laboratory Director: Lew Kowarski (France)
- Director of Theory (TH) Division: Niels Bohr (Denmark)

In June 1952, the CERN prep group met in Copenhagen during a conference and was met with exciting news: the Cosmotron at Brookhaven had just successfully finished its first test runs. The Heisenberg/Bohr group called for immediate construction of a smaller accelerator (Synchro-Cyclotron), insisting on Copenhagen as the construction site. Given that the Auger group favored a more powerful accelerator than the American machine in Brookhaven, the PS group, led by Odd Dahl, was commissioned to construct an accelerator with an energy performance of "up to 20 GeV." Meanwhile it had become obvious that meson research was reasonable only beyond the energy region of 10 GeV.

Plans for the new machine had to be changed considerably since the original design was based on the Cosmotron in Brookhaven. But Odd Dahl loved to be challenged. At the Christian Michelsen Institute (CMI) in the Norwegian mountains, and together with his colleague Frank Goward (who often came over from Harwell in the UK), Dahl began to work on the construction of a machine that could produce more energy for accelerating atomic particles than the machine at Brookhaven. Luckily, the Americans were willing to help. The prolific academic competition between the United States and Europe concerning the energy performance of their machines, and within other fields of nuclear physics—always at eye level and focused on mutual help and support—is and continues to be part of the CERN story.

In August 1952, Odd Dahl, Frank Goward, and Rolf Wideroe—a pioneer in particle physics and a betatron expert—made a trip to Brookhaven to study the new Cosmotron. Milton Stanley Livingston (1905–1986), who had been the driving force behind the construction of the Brookhaven National Laboratory (BNL), met the European delegation with great openness and was even willing to talk about the problems that the project had encountered. At Brookhaven, the Europeans were confronted with the fact that via a new, modified arrangement of magnets called the alternating gradient (AG) principle, the particle beam could be more tightly focused than in the Cosmotron. The Europeans realized that their design for a new accelerator, which was completely based on the Cosmotron, was obsolete. On the other hand, they now knew how its performance could be improved considerably: by focusing the particle beam with better designed magnets. With these preconditions it was even more necessary to cooperate with colleagues who were able to build the magnets to improved specifications, and the British were the only ones who had thorough experience in the construction of magnets for particle accelerators in Europe.

A few months later, from October 4–7, 1952, the third meeting of the CERN committee took place in Amsterdam. The latest developments in accelerator technology—the new AG principle—meant that the PS group had to face even more challenges. The large CERN accelerator, the Proton Synchrotron, now had to be constructed for an increased collision energy of 25 GeV and it had to be done at the same cost. But another problem was finally resolved: in Amsterdam, the question of location was decided upon. Switzerland had offered CERN a large piece of land at Meyrin, a small town near Geneva, as the site for the new laboratory. Geneva has a lot of advantages; the city is centrally located in Europe and can be reached very well from anywhere on the continent. Secondly, Geneva had already been established as a renowned location for international organizations such as the International Committee of the Red Cross, established in 1863; the International Labour Organization; and the United Nations with an imposing Palace of Nations, a complex with 2,800 offices and 34 conference rooms. The property provided to CERN seemed to be an ideal place for the new laboratory. Now it had to be built from scratch and certainly with the new focussing system.

Early in December 1952, CERN Secretary-General Edoardo Amaldi went to England to talk about the next steps of cooperation. In London, at the prestigious

Savile Club, Amaldi met with Sir Ben Lockspeiser, Secretary of the Department of Scientific and Industrial Research (DSIR), who later became a CERN director, and with the young physicist John Adams. Adams, like many of his colleagues, had worked in the radar laboratories of the British Ministry of Aircraft Production during WWII. After the war, precision radar technologies had been transferred to the construction of particle accelerators. Adams and his colleagues had already worked on improving focusing technology and knew that a better focused particle beam would improve the new European accelerator significantly.

Cockcroft and Lockspeiser were quite sure that the key figure in British science policy, Lord Cherwell (i.e., Frederick Lindemann, 1886–1957) would support a collaboration with CERN and that he would back the project politically. To their surprise Lord Cherwell, scientific advisor to the then re-elected Prime Minister Winston Churchill, showed a rather reluctant attitude towards Amaldi's ideas. Lord Cherwell argued that the project was impractical but he nevertheless invited Amaldi to visit the nuclear laboratory at Harwell, where Amaldi wanted to talk to the young British researchers about further planning. Neither Amaldi nor his British colleagues were certain about Lord Cherwell's strong opposition. Perhaps the top league of politicians had decided that it would be wise to vote for an independent British laboratory of nuclear physics instead.

During their three-hour drive out west to Harwell, Amaldi was entertained by his young driver, John Adams. John Bertram Adams (1920–1984) played a decisive role in the construction of the SC at Harwell, but he was now eager to develop his professional capabilities and perform new tasks. Amaldi later recalled that Adams "was remarkable by any standard, and he was ready, incredibly ready, to come to work for CERN." In Harwell, Amaldi met with the new generation of motivated researchers, many of whom were apparently looking forward to future European cooperation. Those young British scientists (e.g., John Adams, Frank Goward, Jim Cassels, and Donald Fry, to name a few) were used to working on difficult, large-scale projects with machines that they had built by themselves. This expertise was what CERN needed and still needs today. Amaldi: "I was also very impressed by the other young people that I met at Harwell, not only because of their scientific abilities but also because of the detailed information they had about CERN." The young British scientists were very interested in the new focusing principle, and how it could be implemented practically. This idea was clearly at the center of their thinking. Adams himself thought that his ideas for the new type of accelerator were an "adventurous high-risk, high-gain course of action."

If successful, the new method would allow for a much more powerful accelerator than the Cosmotron at Brookhaven.

The CERN Convention of June 1953

The sixth CERN Conference started in Paris on June 29, 1953. When the conference came to its conclusion on July 1, 1953, representatives from twelve European States had signed the CERN Convention.

Article 2 of the convention outlines the goals of the organization, which are still valid today:

> The Organization shall provide for collaboration among European States in nuclear research of a pure scientific and fundamental character, and in research essentially related thereto. The Organization shall have no concern with work for military requirements and the results of its experimental and theoretical work shall be published or otherwise made generally available.

The twelve founding members of CERN are Switzerland, France, Belgium, Denmark, Germany, Greece, the United Kingdom, Italy, Yugoslavia, the Netherlands, Norway, and Sweden. When the United Kingdom signed the Convention it ceased to be an observer and instead became a full member state. The Federal Republic of Germany had played a very active part in CERN's founding process (through Werner Heisenberg). With his signature, Germany regained its status as a coequal partner within the family of European nations for the first time since the end of WWII.

To understand the nature of CERN, the relationships between CERN and various national institutions, and its organizational character, the minutes from the CERN Council's seventh session read like this:

> The new laboratory at Geneva should resemble in its research characteristics the research laboratory of a university, through which there is a free flow of 'graduates', and there for [sic] 'most posts' should be of a short duration to allow the flow of scientists through the Laboratory so necessary to avoid stagnation. This will also increase the number of scientists who are able to use the unique facilities of the Laboratory. [...] The Organization shall provide for collaboration among European States of a pure scientific and fundamental character, and in research essentially related thereto.

To realize their ambitious goals, CERN needed—in addition to scientific expertise and enthusiasm—money, lots of money. The financial requirements for a period of seven years were calculated to be 130 million Swiss francs for the construction of laboratories, equipment, management, and maintenance costs. These funds were to be supplied by the Member States...

> ...based on the average net national income at factor cost of each Member State for the three latest preceding years. [...] No Member State shall, in respect of the basic programme [sic], be required to pay contributions in excess of twenty-five per cent of the total amount of contributions. [...] Scale to serve as a basis for the assessment of contributions during the period ending on the 31st of December, 1956: Belgium 4.88 per cent, Denmark 2.48 per cent, France 23.84 per cent, [the] Federal Republic of Germany 17.70 per cent, Greece 0.97 per cent, Italy 10.20 per cent, [the] Netherlands 3.68 per cent, Norway 1.79 per cent, Sweden 4.98 per cent, Switzerland 3.71 per cent, [the] United Kingdom 23.84 per cent, [and] Yugoslavia 1.93 per cent.

By signing the CERN Convention, every participating country made binding commitments. Compared to the first estimates the projected costs of the new laboratory had already risen by 50%. Budget questions had been discussed widely within the CERN Council, especially regarding the issue of whether the budget was too high—an issue that had been debated during every internal and public discussion about CERN up until today. At that early stage, the players were good-humored and looking ahead to the future. The report to the Member States, a paper that was released in Rome in 1953 (March 23, 1953, CERN/32), stated:

> Budget too high? If it is found that the proposed scale of investment and operating costs goes beyond the European ability to pay, we may have to scale down such items as the number of experimental teams, that is the number of experiments simultaneously carried out on either machine; or to curtail some of the supporting theoretical and experimental activities; or to rely to a greater extent on outside institutions for such supporting activities; or to narrow down the teaching and melting-pot scope of the Center. None of these economies is impossible—but if our estimates are to be believed—every one leads to a situation in which a very costly and presumably vary

[sic] promising, fundamental equipment cannot be used to its full capacity of scientific output. It is to be hoped that at the time of decision, those responsible for the translation of our plans into realities will know how to come down to earth without becoming pedestrian.

The CERN schedule for completion of the facilities aimed to conduct first runs of the Synchro-Cyclotron (SC) in 1957. The big machine, the 25 GeV Proton Synchrotron (PS), was scheduled to be completed within seven years, i.e., first runs were expected in 1960. Each of the new machines was scheduled for up to ten experimental teams, with a 15-hour working day for the accelerators.

The Work Begins

Work in Geneva started during the fall of 1953. The 12-men PS team gathered at a site lent by the University of Geneva; other staff had to work from wooden barracks erected close to the Geneva airport (GVA) in Cointrin. Meanwhile John Adams wrote a considerable number of scientific papers regarding the new Alternating Gradient Focusing principle for particle accelerators. Adams could eliminate any doubt about the new technology; now it was time to implement the plans. John and Hildred Blewett, two American physicists who had helped to build the Brookhaven

First excavation work on the Meyrin site. (© 1954 CERN, CERN-HI-5405001)

accelerator, joined the PS group to support the project.

The most important decisions regarding design and technology of the Proton Synchrotron were made by the CERN Parameter Committee, where John Adams, with his clear strategy and precise methodical approach, had the greatest influence. With the help of his congenial British colleague, Mervyn Hine (together they were dubbed, "the terrible twins"), Adams' work sessions often led to significant technological advances.

The Parameter Committee had an atmosphere—to a great extent based on the personality and style of John Adams—in which factual research was enabled by knowledge, know-how, and the drive to keep pushing boundaries further. That spirit characterizes the efficient work at CERN to this day. John Adams' credo was, in his own words:

> The question of how much flexibility to build into a machine is obviously a matter of judgment, and sometimes the machine designers are better judges than the physicists who are anxious to start their research as soon as possible. But whatever compromise is reached about flexibility, one should certainly avoid taking risks with the reliability of the machine because then all its users suffer for as long as it [is] in service and the worst thing of all is to launch [the] accelerator project, irrespective of whether or not one knows how to overcome the technical problems. That is the surest way of ending up with an expensive machine of doubtful reliability, later than was promised, and a physicist community which is thoroughly dissatisfied. CERN, I am glad to say, has avoided this trap and has consistently built machines which operate reliably, are capable of extensive development, and have been constructed within the times promised and within the estimated costs.
>
> — E.J.N. Wilson, *Sir John Adams: His Legacy to the World of Particle Accelerators*, 2009

On May 17, 1954, groundwork began on the CERN site at Meyrin. A few days before, France, as one of the main political forces behind the project, had finally signed the CERN Convention. Where there were now green meadows and arable land, the CERN scientific research facility would be built within the next five years. A tragic death overshadowed those early days; Frank Goward, "the man in charge" of the PS group, who had been responsible for the design and construction of the large accelerator, died in March 1954. Then Odd Dahl, director of the PS

The PS "Parameter Committee" members, 1960. *Left to right*: Ed. Regenstreif, Pierre Germain, Kjell Johnsen, Arnold Schoch, Mervyn Hine, John Adams, Franco Bonaudi, Fritz Grutter, Kees Zilverschoon, and Colin Ramm. (© CERN 2033a, 1960)

group, resigned from his post and went back to his institute in Bergen (Norway) to build the Halden Nuclear Research Reactor. As a result, John Adams, at the age of 34 years, became director of the CERN PS Division. Adams' signature style would continue to mold the CERN scientific spirit for the next 20 years and beyond. Style and efficiency were the hallmark of John Adams and, in general, he was convinced that "international common ventures prevent wars."

Felix Bloch

Felix Bloch (1905–1983) was born in Zurich, Switzerland; he had studied with Wolfgang Pauli in Zurich and with Werner Heisenberg in Leipzig. A brilliant student, Bloch was also lucky enough to receive a one-year scholarship to the Niels Bohr Institute in Copenhagen. After that Bloch went back to Germany, assuming a position as a lecturer at Leipzig University. Coming from a Jewish background, Bloch had to leave Germany in 1933. He immigrated to the United States, where he became the first professor of theoretical physics at Stanford University. During WWII, Bloch worked at the Los Alamos Nuclear Laboratory, later joining the Radar Project at Harvard University. By then he had already made his main discoveries about the properties of neutrons, and in 1946 he worked with nuclear magnetic resonance (NMR). Today, the further development of MRT (Magnetic Resonance

Tomography) is widely used in medicine as an imaging method. Together with his colleague, Edward Mills Purcell, Bloch had won the 1952 Nobel Prize in Physics "for their development of new methods for nuclear magnetic precision measurements and discoveries in connection therewith."

During a meeting in June 1953, proposed by Niels Bohr, the CERN committee had quickly agreed on Felix Bloch as the only candidate to be Director-General of CERN. Bloch was an expert in the new beam focusing principle, thus he was the right man to oversee the construction of the new CERN machine. It was quite clear to the committee that Bloch would only be available for a relatively limited period of time, maybe two years, because Bloch wished to go back to the United States to further his studies at Stanford. It is quite possible that the interim CERN committee had deliberately nominated a candidate with a relatively short "lifetime," so that there would be more time to find the next CERN Director-General from their own ranks.

The Interim CERN Committee was concluded on September 19, 1954. In the meantime the required minimum of seven states had ratified the CERN Convention. Now CERN, born as "Conseil Européen pour la Recherche Nucléaire," needed another name. "Council" would be inappropriate in the long run. The official name was therefore changed to "European Organization for Nuclear Research." The catchy and internationally established acronym "CERN" was kept as it was, which would prove to be a little confusing for later generations.

The Creation of CERN, phases I–III

I. Preparation phase: December 1949 (Lausanne Conference) until February 1952 (Geneva Agreement)
II. Planning phase: February 1952 until July 1953: CERN Convention signed in Paris creates the "Conseil Européen pour la Recherche Nucléaire"
III. Interim phase: July 1953 CERN Convention until September 29, 1954, in Geneva: CERN renamed the "European Organization for Nuclear Research"

Personnel Data

Felix Bloch was appointed Director-General of CERN during the first session of the permanent CERN Council, held October 7–8, 1954, in Geneva. CERN pioneer Edoardo Amaldi was appointed Deputy Director-General. In succession to French delegate Robert Valeur, Sir Ben Lockspeiser, who had previously headed the interim Finance Committee, was elected President of the CERN Council. According to a CERN press release, Lockspeiser was "one of the leading government scientists of Great Britain," bringing to CERN "a wealth of scientific and administrative experience which [would] be of special value at this formative stage." Cornelis Jan Bakker was appointed Director of the Synchro-Cyclotron Division; John Adams, Head of the Proton Synchrotron Group; Lew Kowarski, Head of the Scientific and Technical Services Division; and Christian Moeller, Head of the Theoretical Studies Division. In addition, the CERN Council established three committees: the Committee of the Council (meeting more frequently than the Council itself and dealing with "certain categories of matters of policy"); the Scientific Policy Committee (to "consider and report to the Council on matters of scientific policy," with the Chairman being Werner Heisenberg); and the Finance Committee ("responsible for reviewing the budget", chaired by M. Jean Willems).

The Structure of CERN (as of 1954)

- Director-General and Deputy DG
- CERN Council, with a president and two deputies
- Three committees: the Committee of the Council, the Scientific Policy Committee, and the Finance Committee

---

In his opening speech on November 19, 1954, Felix Bloch, the new Director-General, thanked the initiators of CERN, and especially Louis de Broglie, Edoardo Amaldi, and Isidor Rabi. Bloch emphasized that according to the Convention "the research activities of CERN shall be conducted so as to exclude all military applications." Bloch stated that CERN had no intentions to postpone its research activities until the big machines were completed but that active research was already "beginning in cosmic ray studies and in some other fields." Bloch characterized the main goals of CERN:

> The significance of CERN may be seen not only in the type of research to which it is devoted, but also in the fact that it represents a unique international effort in science. [...] While international collaboration is evidently a desired goal leading to a more peaceful world, it is notoriously difficult to achieve it in many human

activities. If there is one field in which it should be relatively easy to do so, it is certainly that of pure research and the ultimate success of CERN depends greatly upon its taking a leading step in this direction.

On June 10, 1955, Felix Bloch laid the foundation stone on the laboratory site. (© CERN 1955, CERN-HI-5506002)

CERN grew rapidly. On October 1, 1954, the organization had 114 employees already. One month later, there were 180 employees. Felix Bloch admitted that he had underestimated the substantial administrative effort required for the rapidly growing organization. Then the Deputy Director-General of CERN, Edoardo Amaldi, who was quite familiar with all matters concerning the organization, suddenly quit his job, and Bloch asked for his return to the US after just one year in Geneva. This option had obviously been agreed upon with Niels Bohr. On June 10, 1955, Felix Bloch laid the foundation stone of the new laboratory. His successor, however, had already been named; it was Dutch Professor Cornelis Jan Bakker.

The Era of Cornelis Jan Bakker (1955–1960)

> *"CERN represents the most sensible thing Europe has ever produced."*
> — Friedrich Duerrenmatt
> Swiss writer (1921–1990), about his play "The Physicists"

Cornelis Jan Bakker (1904–1960) had been involved in the creation of CERN since 1951. In that year, Bakker was invited by Pierre Auger to act as one of the eight experts assigned to drawing up the plans for the future CERN. In 1952, Bakker was appointed Director of the SC group. At the first session of the Council of CERN, Bakker was appointed Director of the SC Division. In September 1955, he succeeded Felix Bloch as Director-General. During Bakker's directorate, CERN made tremendous progress. The construction site became an enormous complex with rapidly growing buildings, giant assembly halls, and new laboratories and utilities.

CERN will never be completed, CERN will always further develop.

Both in 1956 and 1957, there were extremely harsh winters and extraordinarily long periods of rain in between. At the CERN site, there were many challenges to be met. As a result, the costs skyrocketed. In 1956, the total construction budget was estimated at 40 million Swiss francs. Three years later, in 1957 it was 64 million. In 1956, the total budget for the fiscal years 1952 to 1960 was estimated to be 220 million Swiss francs. Two years before, in 1954 the budget for this period had been estimated to be only 120 million. These financial problems led to some rather radical ideas. Parts of the main building, other administration buildings, the auditorium, and the cafeteria were all considered to be superfluous. If these plans would have been realized, the main hub of CERN, the fabulous CERN cafeteria, would not exist at all. True drama, no?

CERN had difficulties right from the beginning and there were many reasons for it. First, the two new accelerators had long construction and even longer building schedules. It took some time to sort out the details, to develop new construction methods, and to think about and design the components. After all, this was very complex territory. Technological parameters had to be redefined almost constantly, while new and unknown challenges popped up everywhere. Very important staff decisions had to be made. Who should come to CERN in the first place? How should the staff be paid? In which way should the various committees, all of those new departments and tech groups, work together? There were even more complex problems: the Finance Department was always in trouble; it had to deal with rapidly increasing financial demands. And the CERN Scientific Policy Committee, chaired by Werner Heisenberg, had another issue: Copenhagen, headquarters of the Theoretical Studies Division, did not come to fruition.

The problem was that neither Niels Bohr, his assistant Stefan Rozental (1903–1994), nor Christian Moeller, Danish head of the CERN Theoretical Studies Division intended to come to Geneva to work there permanently. During the directorship of Felix Bloch, in 1955, the CERN Council had already stated that the executives of the Theoretical Division were required to work in and out of Geneva:

> It is imperative now to create a working nucleus for a theoretical group in the CERN laboratory. [...] The leading theoretician should be as good a man as can be found [and] he should eventually become director of the Theoretical Division in Geneva.
>
> — CERN, July 25, 1955, CM-P00094772

On the other hand, the Theoretical Division in Copenhagen had given young men the opportunity of training by becoming stipendiaries—but the payment of those young physicists was part of the reason for the resentment. The cost of the Theoretical Studies Division in Copenhagen was one of the largest items in the budget. This issue had to be resolved, and it was solved. The Science Committee, during its sixth meeting on May 15, 1957, elected Bruno Ferretti, professor at the University of Rome, as new Head of the Theoretical Division in Geneva. Ferretti had been involved in the European initiative from its beginnings, making the first concrete proposal to build a large accelerator during a meeting in December 1950. With the establishment of the Theoretical Division in Geneva, business relations between CERN and the Niels Bohr Institute were history.

SC, PS, LINAC, and Other Machines

In 1957, the SC started operations almost exactly on schedule. The SC accelerated protons on a spiral path to a beam energy of 600 MeV. At the time, this CERN machine was third in the world in terms of its power output, close to the machines at Berkeley (USA) and Dubna (USSR). In July 1958, the SC team announced its first discoveries. The CERN scientists had observed the decay of a pion (or pi-meson, a zero spin subatomic particle playing an important role in explaining the properties of a strong nuclear force) into an electron and a neutrino. Since then neutrino research has played an important part in the various fields of research at CERN. Today, this field is labeled "CNGS" (CERN Neutrinos to Gran Sasso). Neutrinos produced by the LHC are transferred to a laboratory deep in the Gran Sasso Massif in Italy. A complete overhaul of the SC in 1973/74 led to an increase of the beam intensity by a factor of four. In 1990, the Synchro-Cyclotron, one of the most reliable CERN workhorses, was finally switched off after 33 years in operation.

After the sudden death of his British colleague Frank Goward in 1954, the CERN Proton Synchrotron Group had been led by John Adams. He masterminded the design and construction of the new machine; it was mainly Adams who brought the technically difficult project to its successful realization. Obviously his main objective at that time was to win the race against the Americans. In 1952, shortly after the interim CERN Council's decision to build a more tightly focused machine using the AG principle, Brookhaven had announced its intentions to build a similar machine, the Alternating Gradient Synchrotron (AGS).

John Adams, November 25, 1959, in the main auditorium at CERN. (© CERN-HI-5901881-1)

The CERN PS is a ring accelerator with a circumference of 628 meters. The machine is composed of one hundred individual magnets with a total weight of 3,800 tons—in contrast to an estimated 800 tons. The construction of the accelerator was completed in 1959, and on July 27 of that year the PS team announced the successful "first run" of the beam. On November 24, 1959, the beam energy reached 24 GeV and in December it was at 28 GeV. It was the world's highest energy proton beam at the time. Today, the PS is still running; the upgraded machine serves as one of the pre-accelerators for the LHC.

It was a high point of his career when on November 25, 1959 John Adams presented his results in the just-completed main auditorium at CERN. In his hand he held a bottle of vodka, which he had received from his colleagues in the USSR. It had come with the message that it was to be drunk only when CERN passed Dubna's world record energy level of 10 GeV. Now the bottle was empty—emptied overnight by John Adams and his colleagues. The bottle was sent back to Dubna the next day, together with the Polaroid photograph of the successful PS run and their very best regards. John Adams' great achievement was celebrated in a press conference on February 5, 1960. More than 110 members of the press attended the presentation, together with Niels Bohr, John Cockcroft, Francis Perrin, J. Robert Oppenheimer, CERN DG Cornelis Bakker, and Edoardo Amaldi.

The start of experiments with the PS was scheduled for spring 1960. Until then, during the design and construction of the machine, any difficulties encountered had been mastered by ingenious team work and great financial resources. Now it was realized that the PS project was hampered by a lack of additional equipment, adequate detectors, and experienced staff. All of this made up for the six-month head start of the CERN PS over the AGS machine at Brookhaven. On top of that, troubles with the slow start of the PS program were overshadowed by the tragic death of Cornelis Jan Bakker, Director-General of CERN since 1955, who died in a plane crash in April 1960. After this tragic event, John Adams was appointed Director-General of CERN for the next 15 months.

The principle of linear acceleration of charged particles by alternating electric fields (AC) was invented by Rolf Wideroe, a Norwegian, in 1928. From 1950–52, Wideroe had worked as a consultant for what was to become CERN. CERN's first proton linear accelerator, LINAC 1 (Linear Particle Accelerator) became operational in 1959. Linear accelerators are used to accelerate charged particles—such as ions—before those particles are injected into bigger systems, such as the CERN PS. In 1978 this task was taken over by LINAC 2. LINAC 1 remained in service until 1992 for further experiments such as the acceleration of oxygen or sulfur ions, and to test RF quadrupole magnets.

ISOLDE, the Isotope Separator On Line DEvice, was commissioned in 1967. This experiment was connected to the SC, which had already been upgraded to 600 MeV. Thanks to the use of improved technologies and new technological ap-proaches, ISOLDE could separate different short-lived radioactive isotopes online. ISOLDE worked with a particle beam that only consisted of single isotopic species. With this technique, ISOLDE opened a new field of research with radioactive ions at CERN. The technology has been further developed up until today, making ISOLDE a world-leading laboratory in the study of radioactive nuclei.

Bubble Chambers, BEBC, Gargamelle

During the 1950s and 1960s, bubble chambers were standard experimental tools in high-energy physics. Bubble chambers are general-purpose devices that can be used in a variety of experiments. A bubble chamber consists of a tank that is filled with a transparent liquid, such as superheated hydrogen. The superheated liquid is placed under pressure (about five atmospheres) and then cooled down to a very low temperature (about three degrees Kelvin). Just before the charged particles hit the cooled liquid, the pressure in the tank is suddenly reduced by expanding its volume using a large piston. When a charged particle flies through the liquid, the energy of the particle is deposited in the liquid and boiling is initiated along its trajectory, leaving a trail of bubbles that can be photographed. According to legend, the American physicist and molecular biologist Donald A. Glaser (1960 Nobel laureate in Physics "for the invention of the bubble chamber") discovered the principle when looking at the bubbles in a full glass of beer. At CERN, the bubble chamber program started in the late 1950s. The Head of the Bubble Chamber Group was Charles Peyrou. Under Peyrou's leadership, and thanks to his exceptional understanding of both physics and engineering, bubble chamber physics saw remarkable progress at CERN. After two years with a ten-centimeter bubble chamber, the HBC (Hydrogen Bubble Chamber) with a diameter of 30 centimeters began operation in 1959.

Experiments with larger bubble chambers began with the HBC 200 in 1964. The chamber was two meters long; more than 40 million photographs were taken over its 12 years in operation. Successful research with the HBC led to the construction of even larger bubble chambers, such as the BEBC and Gargamelle.

Construction plans for the BEBC (Big European Bubble Chamber) were finalized in 1967 by a special financing agreement between CERN, France (SACLAY), and Germany (DESY). The total cost of this experiment amounted to 92 million Swiss francs. The BEBC was huge. It was 3.7 meters in diameter and was equipped with the largest superconducting magnet in the world at the time. Construction work began in 1970 and the first pictures were taken in 1973. The chamber was filled with 35 cubic meters of liquid (hydrogen, deuterium, or a neon-hydrogen mixture); the beam was provided by the 26 MeV PS. During its life, the BEBC facilitated the discovery of D-mesons and furthered the development of neutrino and hadron physics. Until its shutdown in 1984, more than six million photographs had been taken; 3,000 kilometers of film had been exposed; and, during that time, around 600 scientists from more than 50 laboratories had worked successfully with the BEBC experiment.

Gargamelle

Gargamelle was the second large heavy-liquid bubble chamber at CERN; it was built at the École Polytechnique in Paris. The name is derived from Gargamelle, a giantess, and mother of Gargantua in François Rabelais' novel, *Gargantua and*

Pantagruel. The Gargamelle chamber was cylindrical, 4.8 meters long and 1.85 meters wide, with a volume of 12 cubic meters. The system had a conventional magnet, i.e., no superconducting magnet; it generated a magnetic field of 2 T (Tesla). The chamber was filled with 13.5 tons of Freon (CF_3Br), known today as a dangerous climate-killer that was widely used in refrigerators and air-conditioning machines until recently. The relatively "thick" liquid facilitated better observability of the events in neutrino, muon, and pion research. The prime scientific objective of Gargamelle was the observation of neutrino events under the motto "low viscosity for higher probability of neutrino interactions."

In December 1972, one of the greatest discoveries in CERN's research history was made with Gargamelle. Further tests followed before the results were presented at CERN's main auditorium in July 1973. The Gargamelle collaboration announced the discovery of "neutral currents" (or "weak neutral currents"). These currents proved the existence of electroweak interactions, which had been predicted by Sheldon Glashow, Abdus Salam, and Steven Weinberg in 1967. Their theory demanded that electromagnetic and electroweak nuclear forces ("radioactivity") merged in very-high-energy regions. The three physicists won the 1979 Nobel Prize in Physics for their theory. It is one of the pillars of the Standard Model of particle physics.

Much of today's research at CERN is concerned with the Standard Model; it is about finding the elementary (or other subatomic) particles needed to prove the theory. In the original photograph taken at Gargamelle, a neutrino interacts (collides) with an electron. The electron's path is the horizontal line; it leaves a spiral "skid-mark" downwards right after the event. Between 1970 and 1978, approximately 83,000 neutrino events were registered and analyzed. Altogether, 102 neutral current events were observed. At the Gargamelle collaboration, 60 physicists from seven national laboratories were employed. It was the largest European cooperation in this sector of scientific research at its time.

The discovery of neutral currents at CERN was an enormous achievement. Furthermore, the unification of weak and electromagnetic interaction, i.e., the theory of electroweak force (by Weinberg, Salam, and Glashow) necessitated the existence of a hitherto unobserved particle, the so-called "Z boson." The discovery of the "neutral current interaction" by the Gargamelle collaboration verified the existence of the Z boson. From that point on, one of the main research topics at CERN has been the verification of the Standard Model, i.e., finding all of the predicted particles, such as the Z, W, and, finally, the Higgs boson.

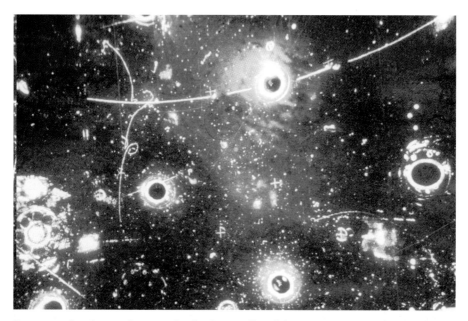

Discovery of weak neutral currents by the Gargamelle collaboration at CERN, 1973. An electron is hit by a neutrino and spirals off. (© CERN 1973, CERN-EX-60100)

CERN bubble chamber collaborations have been extremely beneficial to converging the international research community. Whole generations of physicists were able to work on their theses based on data coming from the CERN bubble chamber experiments. Unique and diverse research opportunities at CERN strengthened ties to other scientific institutes. International cooperation was significantly extended over the following years and decades. Large international collaborations became the cornerstones of the international success story of CERN.

Despite its excellent performance and striking results, Gargamelle was one of the last mechanical bubble chambers. Further research into the mechanisms, forces, and interactions within the atom would have been impossible with bubble chambers because of the numerous disadvantages associated with this detector type. Bubble chambers are not suited for particles with very high energies; the frequency of records (photographs) is far too low, and bubble chambers are too small to track entire trajectories of particles produced at very high energies (particles simply fly further). Today, bubble chambers are used for demonstration purposes only; as detectors, they are irrelevant. However, modern particle detectors basically do nothing else than bubble chambers; they just register events electronically, not mechanically. The improvements in transistor technology—making them ever smaller, and better—boosted the electronic revolution of the 1960s. A mechanical camera connected to a bubble chamber was able to deliver about one image per second. An electronic detector could take many more shots and could also be far more specific.

Georges Charpak's Multi-Wire Proportional Chamber (MWPC)

Georges Charpak's multi-wire proportional chamber. (© CERN 1973, CERN-EX-7304218)

Georges Charpak (1924–2010), a former resistance fighter and survivor of the Dachau concentration camp, was employed at CERN from 1959. In the early 1960s he developed a variety of non-photographic measurement methods. In 1968, he eventually built the "multi-wire proportional chamber" (MWPC). Charpak's new detector was a gas-filled box wrapped with a large number of parallel wires. Each one of those wires was connected to an amplifier. In cooperation with a computer, the camera could record up to one million particle tracks per second, a huge improvement over the previously used mechanical bubble chambers. Charpak's electronic particle detector revolutionized the way elementary particles and their decay products could be detected. Today, every modern experiment in particle physics—and in many other areas of science, such as biology, radiology, and nuclear medicine—is working with detectors based on the principle of Charpak's multi-wire proportional chamber. The 1992 Nobel Prize in Physics was awarded to him "for his invention and development of particle detectors, in particular the multi-wire proportional chamber."

The ISR and Other New Proton Accelerators

After the completion of the Proton Synchrotron in 1960, the PS group under John Adams started thinking about the next generation of accelerators. Unfortunately, Adams went back to England in 1961 to explore the feasibility of a fusion reactor (Project "ZETA"). His successor as Director-General, Victor F. Weisskopf

(1908–2002), continued planning and construction of new and larger accelerators. In 1965, the CERN Council voted for two new projects:

1) Intersecting Storage Rings (ISR): The first proton-proton (hadron) accelerator in the world, using protons with an operating energy of up to 28 GeV.
2) A new proton accelerator (Super Proton Synchrotron, SPS), which would have an operating energy of up to 300 GeV (the "300 GeV project").

CERN expenditures continuously rose but the CERN member states had continuously reduced their budgets for science and research since the 1950s, from up to 15% to a few percentage points of the national budget. The ISR had a budget of 312 million Swiss francs; the even bigger 300 GeV project, with an estimated budget of 1.5 billion Swiss francs, was thus not decided upon until 1971.

To facilitate the construction of the ISR, the CERN Council signed an agreement with France. The existing site in Switzerland was expanded across the French border, where the ISR could be built on 40 acres of land. The ISR was designed after a completely new principle. Until then, accelerated particles had been collided with a solid, unmoving target. In contrast, the ISR consisted of two concentric rings of magnets 300 meters in diameter, in which protons travel in opposite directions and are brought to collision. Construction of the underground tunnels began in November 1966. The ISR had a circumference of 943 meters, exactly 1.5 times larger than the PS. Design pressure inside the beam pipe was about 10 torr (1 atm = 760 torr), a very high vacuum for that time. The ISR was at the forefront in many technological fields and a milestone in the history of particle accelerators; it became the new "workhorse" of CERN.

Aerial view of the ISR site. (CERN-PHOTO-7808537X)

The first proton-proton collisions in the ISR were recorded on January 27, 1971. Even in the first tests the overall performance of the whole system was quite respectable—the new machine operated with a remarkable availability of 95% in its first year. In the following years, the ISR proved to be extremely reliable and accurate. Test series of up to 60 hours (later up to 300 hours) operation time were performed. Beginning in November 1980, superconducting magnets were used in the ISR for the first time; they increased its luminosity (i.e., event density) even further. The machine was in operation until 1984. Many technical challenges have been solved with the experimental work done at the ISR, for instance, in the fields of vacuum technology and stochastic cooling.

From PS to SPS — the Return of John Adams

John Adams served as director of the Culham Fusion Laboratory from 1961 through 1966. Until 1971, he worked in an executive position in the United Kingdom's Atomic Energy Authority. At the end of 1969, Adams came back to Geneva to become head of the follow-up project for the PS, the 300 GeV project. The calculated cost of the new "Ueber-machine" was in excess of one billion Swiss francs—tenfold the budget of its predecessor, the Proton Synchrotron.

The 300 GeV project called for the construction of a completely new machine that had to be significantly larger than the PS to achieve the desired energy horizon. Construction of the large super-machine had been delayed by several years already because of its exhorbitant cost. In the meantime, many technological concepts and layouts were discussed, along with several possible sites in Europe. Such a machine could be built anywhere from scratch, not just at CERN in Geneva. Finally, in December 1970, John Adams proposed the plan that the new laboratory could be built in France, close to Swiss-side CERN, using the existing accelerators LINAC and PS as pre-accelerators. This meant that an even higher level of energy of up to 400 GeV could be achieved with the new machine and that the budget would be even lower than originally anticipated. This principle of using the existing accelerators and adding new, larger accelerators with higher particle energies made John Adams the "father of the giant particle accelerators" that have made CERN a leader in the field of high-energy physics.

In 1971 the Super Proton Synchrotron was finally approved by the CERN member states. Designed by John Adams and his team, the new plant was, logistically and technologically, new territory. The SPS was to be built in a tunnel 40 meters underground with a circumference of approximately seven miles (6,912 meters). Even the CERN Convention had to be changed to make construction of the underground tunnel system possible. Legally, the CERN Convention was applicable only to one single laboratory. The huge size of the new SPS accelerator ring made it necessary to establish a new, independent site in Prévessin (Pays de Gex, France). The two laboratories, each having its own administrative structures and general managers, were merged again in 1976.

Two years later, on July 31, 1974, the huge tunnel-boring machine had finished the tunnel, crossing the Swiss-French border twice underground. During the following two years, the SPS was built, composed of almost one thousand magnets. On June 17, 1976 John Adams stood in front of the members of the CERN Council, reporting the first beams were operating at an energy level of 400 GeV. At the end of 1978, an operating energy of 450 GeV was reached. Once the SPS was in operation, John Adams went on to again become the Director-General of

Simon van der Meer and Carlo Rubbia (left) celebrate their Nobel Prize in 1984 with a toast at CERN. (© CERN 1984, CERN-HI-8410523)

CERN from 1976 to 1980. During that time, he was especially concerned with designing the next big machine—and securing the funding for it.

From 1979 to 1981, the SPS was converted into a proton-antiproton accelerator to be operated after the collider principle (two particle beams circulating in opposite directions), which had already been successfully tested with the ISR. The collider principle is the basic structure of modern accelerators. This enabled experiments that could look deeper into the inner structure of the protons. The new SPS was labeled "SppS"; two detector complexes, Underground Area 1 (UA1) and Underground Area 2 (UA2), were integrated. They recorded the first proton-antiproton collisions in July 1981.

In 1983 the two-detector experiments conducted by a team led by Carlo Rubbia (UA1 and UA2), discovered the W boson (January) and the Z boson (May). By proving the existence of those gauge bosons—particles that carry one of the four fundamental forces, the weak interaction—the Standard Theory of Electroweak Interactions (by Glashow, Salam, and Weinberg) was substantiated along with the basic structure of the Standard Model of particle physics itself. The 1984 Nobel Prize in Physics was jointly awarded to Carlo Rubbia and Simon van der Meer "for their decisive contributions to the large project, which led to the discovery of the field particles W and Z, communicators of weak interaction" (see interview with Carlo Rubbia).

From Large Electron-Positron Collider (LEP) to Large Hadron Collider (LHC)

After their discovery, the exact properties of the W and Z gauge bosons had to be further studied. A new accelerator needed to be built that would be capable of colliding electrons and positrons instead of protons and anti-protons to produce large numbers of gauge bosons. The new method had one major advantage, the fact that electrons and positrons are point-like particles. Therefore electron-positron collisions at very high energies produce symmetrical events without too much ambiguity. In contrast, collisions with protons and anti-protons—particles composed of other, smaller particles—result in "dirty" collisions. In a paper issued by the CERN Science Policy Committee in May 1979, the requirements for the next accelerator project were defined as such:

> The basic reason for choosing an electron machine is this. In the past decade we have learned that protons are essentially composite particles, made up from three subnuclear objects called quarks. This leaves us with the electron as the only elementary particle available for acceleration to high energies. If one wants to probe at small distances, and therefore with very high energies, one must use not bricks, nor atoms, nor nuclei, nor even protons, but electrons.
>
> (CERN/SPC/435/Rev.)

CERN's direction in research was closely linked to competing projects in the United States (at Fermilab and Brookhaven) and the USSR (Dubna) as well. The difference was that at CERN—the largest particle physics research laboratory in the world—the focus of research was increasingly placed on the properties of the vector bosons (the W and Z bosons). Already at that time, another elusive particle was on top of the shopping list for CERN physicists: the Higgs boson. The existence of this particle had been postulated by the Scottish physicist Peter Higgs in 1964—at the same time as François Englert, a Belgian theoretical physicist, and five other particle physicists. This (then still unknown) particle was said to be responsible for the mass of all known particles. In 1979, CERN physicists stated very specific and targeted views about the Higgs boson and a few other fundamental questions:

If the Higgs particle (or whatever mass generating mechanism replaces it) still remains obscure, then we will have to go to higher energies. [...] Major advances in science have always led to the identification of more profound scientific questions and we can with confidence state that throughout the history of particle physics it has never before been possible to identify so readily the scientific questions that need to be resolved and also what accelerators need to be built to enable the crucial experiments to be done. [...] If Europe could start the construction of a large electron machine at the beginning of the next decade we would be, by the end of that decade, in a very strong position. While competition cannot be our main motivation, certainly it must be allowed to play a role, and we see an excellent opportunity for Europe to be first in this domain of human activity.

(Proposal for the next major accelerator project
for CERN, CERN/SPC/446)

CERN was a global player now with an annual budget of 600 million Swiss francs; building the Large Electron-Positron Collider (LEP) would swallow most of it. Drawing up the plans for the future CERN flagship machine began in 1975. They were repeatedly reworked, always seeking a balance between the demands of the scientific community and its feasibility. The biggest challenge was the construction of the underground tunnel, which had to be cut through rock to house the accelerator facility. The tunnel had to be quite long to further accelerate the particles needed to produce the W and Z bosons. The problem was that the more the trajectory of high-energy electrons is bent, the more radiation they emit—they simply lose energy through "synchrotron radiation." Vice versa, it is also true that the greater the radius of the accelerator, the lower the energy loss. According to different calculations the new tunnel had to be larger than anything else before. The original plans envisioned a tunnel with a circumference of 52 kilometers; the final agreement foresaw only 27 kilometers. The technological challenge was enormous because the tunnel had to be drilled at an average depth of 100 meters below the surface and with a deviation of less than a centimeter over the entire length.

The excavation of the LEP tunnel, situated between the Jura Mountains and Lake Geneva, began with an official ceremony in the presence of the representatives of France, François Mitterrand, and Pierre Aubert of Switzerland on September 13, 1983. The project was set back by several months due to flooding but the ring tunnel was finally completed on February 8, 1988. The excavation of the LEP tunnel was Europe's largest civil engineering project prior to the Channel Tunnel. Less than half of the 1.4 million cubic meters of excavated material came from the tunnel itself. The caverns where the experiments would be located, numerous galleries, transfer tunnels for the connection with the pre-accelerators, and several giant vertical service shafts accounted for most of the underground space needed for the technological masterpiece. CERN justified the endeavor, associated with the need for scientific evidence and gigantic experiments, with the principle:

Theory may be the backbone of physics, but experiments are its walls.

The LEP was another milestone in the history of experimental physics at CERN. Between 1989 and 1993, the four LEP detectors (ALEPH, DELPHI, L3, and OPAL) analyzed the production and decay processes of more than 10 million Z bosons. In 1995, the LEP was upgraded for a second operation phase, with the addition of as many as 288 superconducting magnets. This doubled the particle energy that allowed for the production of W^+ and W^- pairs, the other two vector bosons of the weak interaction. After 11 years of successful research, the LEP was decommissioned on November 13, 2000. The entire system was removed from the tunnel and caverns to make way for the construction of a new "super machine" in the same tunnel. The Large Hadron Collider was designed to venture into even higher energy regimes to facilitate the discovery of the yet undetected Higgs boson that had been postulated by Peter Higgs and his colleagues.

The LEP experiments allowed for the precise measurements of many particles of the Standard Model. The extraordinary precision of the values confirmed the model on a solid experimental basis and raised a whole new set of questions, triggered by observations of other cosmic phenomena:

- Does the Higgs boson—or, better yet, the Higgs mechanism—exist?
- Why do elementary particles have such extremely different masses even if they belong to the same "families"?
- Why is there so much more matter than antimatter in the universe?
- What is dark matter? What is it made of ?
- What causes the universe to expand? If it is dark energy, what is it made of ?
- Are there extra dimensions? If so, where are they?
- Are the observed elementary particles really fundamental particles or do smaller components exist?

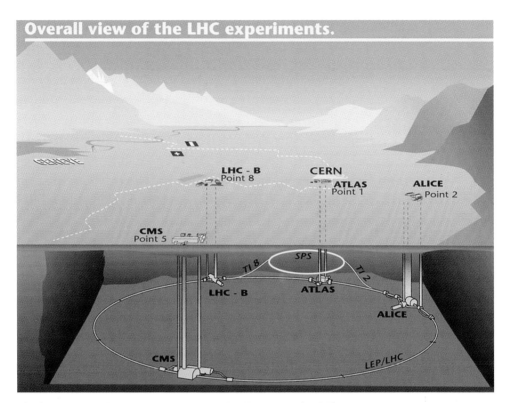

The four main LHC experiments: ALICE, ATLAS, CMS, and LHCb.
(© CERN 1999, CERN-AC-9906026)

- Are there any super-symmetric particles—the much heavier "sisters" of the known particles—and would they be candidates for dark matter particles?

Based on data recorded during the last months of the scheduled runtime of the LEP experiments, the mass of the Higgs particle was estimated at around 115 GeV. The Tevatron in the US had made similar measurements but with an even lower confidence level—not enough by far to claim a discovery. If particles with higher mass were to be detected, and answers for many of the other still open questions were to be found, one had to build a new, much more powerful machine. The high-energy regions where heavy particles (Higgs and SUSY particles) could be detected were impossible to create with electron-positron collisions. However, protons from positively charged hydrogen nuclei lose less energy in the "roundabout" of ring accelerators than electrons. Therefore protons, which are hadrons, can be accelerated more easily—hence the name of the new super machine at CERN: the Large Hadron Collider.

The Large Hadron Collider

The concept for the new particle accelerator and its specifications had been discussed at CERN since the early 1980s. Originally, the new machine had been designed for proton-proton collisions of up to 10 TeV (teraelectron volts = 1 trillion electron volts) of kinetic energy but the first proposal was rejected by the CERN Council in December 1993. After changing the design of the LHC to a collision energy of $2 \times 7 = 14$ TeV, and after some initial financial difficulties had been solved (Germany, the largest contributor to CERN, had reduced its contribution substantially over the years), the LHC was finally commissioned by the CERN Council on December 16, 1994. The Large Hadron Collider (circumference: 26,659 meters, radius: 4,243 meters) had an estimated budget of 2.6 billion Swiss francs. It is the world's largest and most powerful particle accelerator. The US competitor, the SSC (Superconducting Super Collider), a giant machine 87 kilometers in circumference, had been canceled a year earlier after 2.2 billion US dollars had been spent on it. The United States was granted observer status to be further involved in high-energy research at CERN after a significant contribution of about 530 million US dollars.

Initially, the tight budget implied that the LHC was to be conceived as a two-phase project but after substantial contributions from the US, Japan, India, Russia, and Canada, the CERN Council voted in 1995 to allow the project to proceed in a single phase. More than 100 countries contributed to the construction of this gigantic machine. In the end, it took more than ten years to build the LHC and the final budget is—officially—in the range of 5 billion Swiss francs.

The civil engineering works for the LHC Project were divided into three main packages:

1) Excavation and enlargement of the already existing caverns of the LEP tunnel for the experiments ATLAS, CMS, ALICE, and LHCb, plus construction of the necessary vertical shafts. The LHC was financed entirely by CERN. The experimental collaborations are funded independently; CERN contributes to their budgets at a level of 14–20%. The rest had to be raised by the collaborators.
2) Design, construction, and equipment required for the experiments, control centers, and service facilities.
3) Design, construction, and installation of the accelerator, with 1,232 superconducting dipole magnets plus approximately 8,500 magnets of different types for focusing and particle beam guidance.

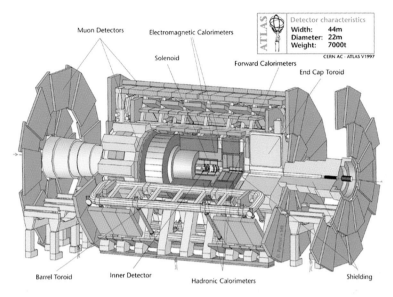

Layout of ATLAS (A Toroidal LHC Apparatus), the largest particle detector ever built. Multipurpose facility for the detection of the Higgs particle; supersymmetric particles and study of CP violation (see interviews with White and Butterworth). (© CERN 1998, CERN-DI-9803026)

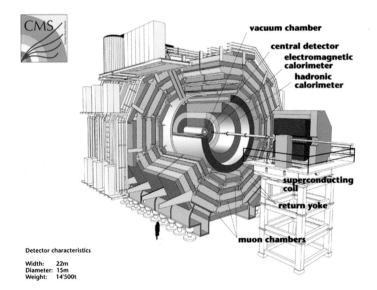

Layout of the CMS (Compact Muon Solenoid) experiment, a general-purpose detector to investigate a wide range of physical phenomena (the Higgs boson and SUSY). It was created in competition with the ATLAS experiment but with a different methodology (see interviews with Virdee and De Roeck-White). (© CERN 1998, CERN-DI-9803027)

Construction of the underground caverns and the access and supply shafts for the ATLAS experiment started in 2001. Starting in 2003, the detector components were shipped to CERN and assembled in the ATLAS experiment pit after being lowered through the vertical shaft, which is 70 meters long and 16 meters wide. ATLAS is the largest collider-detector ever built; the ATLAS cavern is 35 × 40 × 55 meters. The general-purpose detector, designed to cover the widest possible range of elementary physics, has a total weight of 7,000 tons, which is approximately as much as the Eiffel Tower in Paris.

The building engineers of the CMS experiment encountered unexpected difficulties. Their excavation work had to be stopped for two years when they came across the remains of a Gallo-Roman villa complex and archaeologists had to be given priority on the site. In addition, the CMS cavern was directly above a groundwater aquifer that had to be sealed. In contrast to the original plans, the components of the very heavy CMS detector (dimensions: 21 × 10 × 13 meters, weight: 12,500 tons) had to be assembled above ground and then lowered into the cave through the huge supply shaft. Despite the difficulties in building and financing the CMS, ATLAS, and other experiments (the LHCb, ALICE, TOTEM and

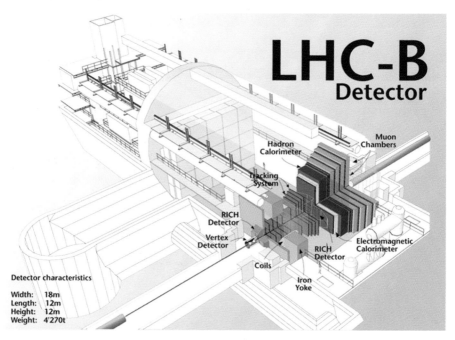

Layout of the LHCb detector. The main purpose of the LHCb detector ("b" stands for a heavy quark known as "bottom") is to investigate nature's preference for matter over antimatter ("CP violation") (see interview with Tara Shears).

The LHC dipole magnet. Sample displayed in front of the CERN cafeteria. (© Michael Krause)

LHCf), they were all completed as planned after almost ten years of construction in spring 2008.

Building the LHC required more than 1,200 superconducting dipole magnets. Each dipole is 14.3 meters long and weighs around 35 tons. They were produced by companies in France, Germany, and Italy according to the contract commissioned by the CERN Council in 2002. These magnets have a hefty price tag of approximately 700,000 euros each; they are precision instruments in themselves.

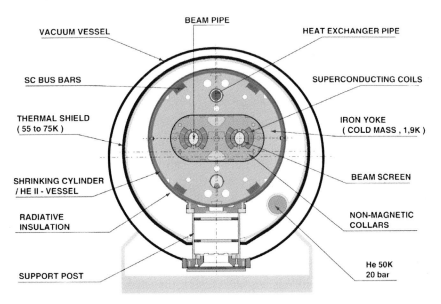

Cross Section of LHC Dipole

The cross section of an LHC superconducting dipole magnet. (© CERN 1996, CERN-AC-9602021 02)

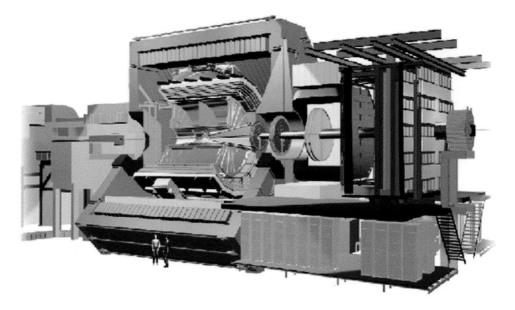

Layout of the ALICE detector. The ALICE (A Large Ion Collider Experiment) experiment studies the collisions of lead nuclei to produce a state of matter known as "quark-gluon plasma". (© CERN 2003, CERN-EX-0307012)

The magnets accommodate two tubes side by side (with a diameter of 56 millimeters each and a distance of 194 millimeters) in which the protons circulate in opposite directions. The particle beam has a thickness of about 1 millimeter; as the beams approach the collision points, they are squeezed to about 16 micrometers, i.e., half the width of a human hair.

The LHC is the largest cryogenic system in the world. 10,000 tons of liquid nitrogen and 120 tons of liquid helium are used to cool down the tubes to −271.3 degrees Celsius or 1.9 degrees Kelvin. This temperature is required to operate the superconducting magnets. An operating current of 11,850 amperes flows in the dipoles to create a magnetic field of 8.3 Tesla, which is 10,000 times stronger than Earth's magnetic field. The internal pressure of the LHC is 10^{-13} atmospheres, ten times less than the pressure on the Moon.

The last dipole magnet was installed in 2007. The following tests over a three-kilometer long test section demonstrated the excellent performance of the system. There are a large variety of magnets in the LHC: dipoles, octupoles, etc.; altogether more than 50 different designs exist, yielding a total of about 9,600 magnets. They accelerate, bend, and focus the two beams. The protons in the beam make 11,245 circuits every second; they travel at 99.9999991% of the speed of light.

CERN Particle Accelerators

At CERN in Geneva various types of particle accelerators are in use today. The protons needed for the experiments start their journey in a tank of research grade hydrogen gas. The molecular hydrogen is broken down into atomic hydrogen or individual atoms. Next, hydrogen's lone electron is stripped from the atom so that a sample of pure protons is left. This happens in the "Duoplasmatron," a device created by the German inventor Manfred von Ardenne in the late 1940s. Next, the protons are passed into a linear accelerator (LINAC2). The LINAC accelerates the protons to an operating energy of 50 MeV. After that the protons are injected into the Proton Synchrotron Booster (PSB), the Proton Synchrotron, and the Super Proton Synchrotron. The SPS accelerates the protons to an operating energy of 450 GeV. At this stage, the particles travel with a velocity of 99.9998% the speed of light. The SPS then injects the protons into the Large Hadron Collider. The LHC accelerates the counter-circulating beams of protons from 450 GeV to

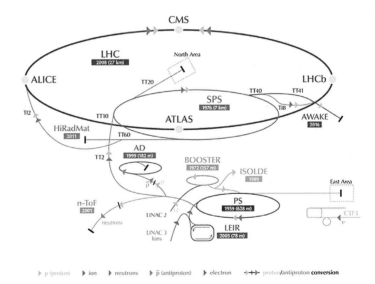

The CERN accelerator complex. The LHC is the last ring (dark grey line) in a complex chain of particle accelerators. The smaller machines are used in a chain to help boost the particles to their final operating energies and provide beams to a whole set of smaller experiments. (© CERN OPEN-PHO-ACCEL-2013-056-1)

4 TeV (in the final stage up to 7 TeV). The protons collide at four collision points where the six experiments (ATLAS, CMS, ALICE, LHCb, LHCf, and TOTEM) are installed. The LHC collides protons (hydrogen nuclei) and lead (Pb) ions as well—the much heavier lead ions have a collision energy of up to 1150 TeV. The other CERN accelerators are: Antiproton Decelerator (AD), ISOLDE, and CNGS. LINAC4, SPL (Super Proton LINAC), and PS2 are all currently under construction.

The project leader of the LHC was Dr. Lyn (Lyndon Rees) Evans, a Welsh scientist born in 1945. Dr. Evans has been working at CERN since 1969. After working on the SPS and LEP, Lyn Evans became involved in the planning of the project (see interview with Lyn Evans).

The Large Hadron Collider is the largest scientific instrument ever built. Its technology is breathtaking and at times almost unimaginable. The protons of the LHC circulate around the ring in well-defined bunches. Under operating conditions each beam has 2,808 bunches, with each bunch containing about 10^{11} (100 billion) protons. The LHC uses a bunch spacing of 25 nanoseconds, which translates to intervals of 7.5 meters. Traveling at near-light speed, a proton in the LHC makes 11,245 circuits every second. Bunches cross on average about 30 million times per second; thus, the LHC generates up to 600 million collisions per second. However, after filtering there are only about 200 collisions of interest per second. In addition to these interesting events, a lot of "background" is created—data from all the other uninteresting events that need to be filtered out in a very complex computing process. The requirements for recording and selecting the results are enormous. This led to the conclusion that this enormous amount of data must be processed by thousands of scientists around the world on a global computing grid.

Types of Particle Accelerators

Particle accelerators are devices that are used to accelerate charged particles (electrons or ions) to very high energies. They are the largest and most expensive pieces of equipment with which physicists do their research to pin down the structure and constituents of matter. During the past decades, ever more powerful accelerators have been built due to the fact that the higher the energy level produced, the deeper one can look into atoms.

All particle accelerators have three basic components:

• A source of elementary particles or ions

- A tube in which particles are accelerated in a high vacuum
- Apparatus (magnets) to accelerate, bend, and focus the particles

    There are two classes of particle accelerators:
- Electrostatic accelerators use static electric fields to accelerate particles. The best known examples are the cathode ray tube from an old tube television, the Cockcroft-Walton generator, and the Van de Graaf generator. At high voltages (breakdown voltage is 15 MV), however, the magnetic field collapses.
- Other accelerators work with high-frequency alternating currents in the radio or microwave range in order to accelerate the particles. Leo Szilard, Rolf Wideroe, and Ernest Lawrence were the pioneers in this field. They built the first large accelerators: the Betatron and the Cyclotron.

LINACs

Linear particle accelerators propel particles by subjecting the charged particles to a series of oscillating electric potentials along a linear beam-line. The largest linear accelerator is the two-mile long Stanford Linear Accelerator Center (SLAC) in Menlo Park, California. It has been operational since 1966 and accelerates electrons or positrons to an operating energy of up to 50 GeV.

Cyclotrons

The cyclotron was invented by the American physicist Ernest Orlando Lawrence (1901–1958) to reduce the large space requirements of a linear accelerator in 1929. Cyclotrons accelerate particles through high-frequency electromagnetic fields (HF-AC) that are applied between two D-shaped electrodes ("dees"). The charged particles are injected near the center of the magnetic field and they travel outwards along a spiral path, increasing their energy (speed) as they go. A cyclotron accelerates particles up to 80% of the speed of light. Cyclotrons were the best source of high-energy particles for experiments for several decades.

In 1941, the cyclotron was used to generate mesons—particles already known from experiments with cosmic rays. Ernest Lawrence was awarded the 1939 Nobel Prize in Physics "for the invention and development of the cyclotron and for results obtained with it, especially with regard to artificial radioactive elements." Lawrence also explored applications for the machine in medicine and biology. In tribute to him, the chemical element with the atomic number 103 was named Lawrencium.

Synchrotrons

The synchrotron is a circular accelerator working with a fixed target. It is the first accelerator type to enable the construction of large-scale facilities because accelerating, bending, and focusing the beam are all done with separate components. In a synchrotron, the charged particles are accelerated in bunches traveling in evacuated pipes through a synchronized, high-frequency electrical field. Synchrotrons reach much higher energy levels than cyclotrons; they can accelerate particles up to almost the speed of light. The concept of the synchrotron was developed by Vladimir Veklser in Russia in 1944 and in the same year by Edwin MacMillan in the United States at Brookhaven National Laboratory. The first synchrotron, the "Cosmotron," recorded collisions in 1952. Further development of the principle (through Alternating Gradient focusing with the PS and SPS at CERN) made the synchrotron a very successful accelerator. Today, electron synchrotrons are also very helpful tools in medical applications and materials research.

Storage Rings

Storage rings are collision machines where particles are accelerated in counter-circulating beams to almost the speed of light. Technically, the geometry of these machines is the same as in a synchrotron. Storage rings were a huge technological step in the development of particle accelerators because the energy of particles colliding head-on can be much higher than in other types of accelerators, and the machine may hold the particles for hours. A stringent requirement for storage rings is a much better vacuum than that found in a synchrotron in order to avoid collision with any residual gas in the pipe. The first electron-positron storage rings were built in the early 1960s in Italy at the Istituto Nazionale in Frascati and in Russia. The first hadron (proton-antiproton) accelerator in the world, the ISR, was built at CERN in 1971. Stochastic cooling and the use of superconducting magnets made hadron colliders such as the LHC the most successful machines in high-energy particle physics.

The CERN Data Center, the World Wide Web, and Beyond

CERN was always in need of high computing power. In 1958, a Ferranti Mercury Computer was installed; it was a state-of-the-art machine that came with a price tag of 980,000 Swiss francs but was actually 1000 times less powerful than a modern PC. The new computing machine filled an entire room but its total computing power would not even be enough to record a single proton-proton collision from the LHC. One year later it was already clear that a more powerful system was needed to deal with the streams of data coming from the experiments. More modern items arrived at the computing center in the form of an IBM 709 in January 1961. The IBM machine—it had a list price of 10 million Swiss francs—was still working with tubes but it was 4–5 times faster than the Mercury; it featured FORTRAN, a program still in use today, and a core memory size of 32K. Soon the first generation of smaller computers (IBM and HP) arrived. By the end of the 1960s, 50 of the then so-called "minis" were in use at CERN. The rapid development of electronics in the 1960s and 1970s made ever higher processing power possible to keep up with the rapidly growing amount of data produced by the experiments.

More processing power was not the only thing in demand; improved communication and information management were necessary, too. The "bottleneck" came from the fact that all of the existing computers (at CERN and elsewhere in the world) could only be addressed separately because the trend towards decentralization and the separation of functions had irreversibly started. To make communication between the experiments and the particle physicists scattered all over the world easier and more effective, Sir Timothy John "Tim" Berners-Lee, a British computer scientist working at CERN, developed a new networking system that eventually became the "World Wide Web" (WWW) in 1989. During the development period, its basic concepts such as URL, HTTP, and HTML were defined. They are still in use today. The first browser and server software was also created at this time. Berners-Lee made his ideas freely available with no patent and no royalties due. CERN management decided to give this invention to the "public domain" and make it available to everyone for free. The Internet, a global network of information and easy accessibility that was previously used by only a handful of scientists, military personnel, and computer nerds, was made available to the general public in this act of generosity.

Tim Berners-Lee and his Belgian colleague, Robert Cailliau, designed the simple system that connects all of humanity today. The Net has revolutionized free access to information and generated whole new industries very quickly. It is still changing the world, maybe more than any other invention created at CERN; today, more than 2.2 billion people use it (as of December 31, 2011). In 2004, Berners-Lee was awarded the Millennium Technology Prize for inventing the World Wide Web. The cash prize was worth 1 million euros. In the same year he received a knighthood when he was promoted to Knight Commander of the Order of the British Empire (KBE) by Queen Elizabeth II. Today, Dr. Berners-Lee is the director of the World Wide Web Consortium (W3C), an organization whose mission is "to develop the full potential of the World Wide Web." Sir Berners-Lee has continued to emphasize the crucial role of CERN in the development of the Internet. Regarding his work, TIME magazine wrote, "He wove the World Wide Web and created a mass medium for the 21st century. The World Wide Web is Berners-Lee's alone. He designed it. He loosed it on the world. And he more than anyone else has fought to keep it open, nonproprietary and free." (*TIME*, 05/17/2010).

The idea of the LHC Computing Grid (LCG) came up in 2005. It is designed to link as many computers together as possible to ensure that the huge amount of data provided by the experiments at CERN can be processed via a common computer platform. The main objective of the LCG is to filter out relevant events from the huge amount of data for further research. In fact, the LHC experiments produce about 1% of the total amount of global data. Each year, approximately 25 petabytes (= 25×10^{15} bytes) are produced by the LHC. The CERN Data Center operates a PC farm with about 9,000 parallel servers representing around 20% of the required total capacity—the power consumption of this computing monster is 3.5 MW. The LCG is connected to 170 computing centers in 36 countries. Data delivered by the LHC and its experiments can only be adequately evaluated with the help of this computing grid. It is a model of what computing will look like in the future; more complex operations are no longer expected to be handled on-site, but through cloud computing.

The LHC Incident of 2008 and Restart

On September 1, 2008, the LHC project leader, Lyn Evans, started operation of the Large Hadron Collider. All of the tests went fine until on September 19 an incident in sector 3-4 shut the accelerator down. The cause of the incident was a faulty electrical connection between two of the accelerator's magnets. This resulted in a high-energy electric arc, which led to massive mechanical damage of the surrounding magnets and a release of several tons of helium coolant into the tunnel. The LHC tunnel had to be closed for a month to let the temperatures of the magnets in question rise close to room temperature before repairs could be started. A total of 53 magnets were badly damaged and had to be replaced. To avoid such incidents in the future, a new monitoring and warning system has been installed. The incident delayed the start of operations by 14 months.

On October 23, 2009, the LHC was restarted. After the unforeseen incident, current Director-General of CERN, Rolf-Dieter Heuer (see interview), limited the machine's performance to half-power, at 3.5 TeV per proton beam. This operating energy was reached on March 30, 2010; since then the collision energy has been increased to 4 TeV. The LHC has been more reliable and accurate than its designers, theorists, or experimentalists at CERN expected. The big machine consistently delivers a higher number of events than planned. After a rebuilding phase in 2014–15, the LHC will be started again to generate collision energies of up to 7 TeV per particle.

Geneva, July 4, 2012

A seminar was held at CERN on July 4, 2012—Independence Day for Americans. It turned out to be the most important event in particle physics in decades. ATLAS experiment spokesperson Fabiola Gianotti said that observations had clearly found signs of a new particle at the 5-sigma level in the energy region of around 125 GeV. Of course, the results were said to be preliminary but a 5-sigma level of confidence is quite dramatic. This means that the discovery is genuine and that a new particle, which is definitely the Higgs boson, had been found. This discovery is a milestone. More complete details of the whole picture are still eagerly awaited. The next steps will be to determine the precise nature of the new particle and what it means for our understanding of the universe.

CERN Today

CERN's main campus is located at Meyrin in the northwest of Geneva in Switzerland on the Franco-Swiss border. CERN is the world's largest particle physics laboratory. The organization operates a network of accelerators. Each machine increases the particles' energy before delivering them to experiments or to the next, more powerful, accelerator. Most of CERN's activities are directed towards the underground Large Hadron Collider (LHC) and its related experiments

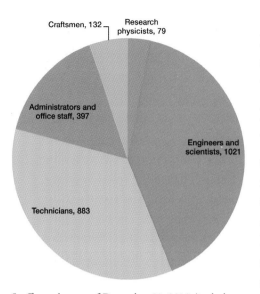

Staff members as of December 31, 2012 (includes externally funded individuals): 2,512.

(e.g., ATLAS, CMS, LHCb, and ALICE). Large parts of the accelerator rings and some experiments are located on French territory. The French side is under Swiss jurisdiction and there is no obvious border within the site. CERN is administered by the CERN Council. Member states appoint two delegates to the Council, a representative of the government and a scientist who together have one vote. The official working languages are English and French. CERN has about 2,500 full-time employees; 1,500 part-time employees; and around 10,000 visiting scientists representing more than 600 universities and 113 nationalities. CERN's annual budget in 2013 was 1.24 billion Swiss francs.

A question regarding CERN that comes up often is: what are the benefits? CERN is a huge think tank with worldwide impact in many scientific fields. Fifty percent of the physicists concerned with fundamental research are employed at CERN. Their research pushes the boundaries of knowledge forward. This is beneficial for the development of new technologies and their fields of application worldwide. The Internet was born at CERN and it has revolutionized our daily lives, our knowledge, and our telecommunication. CERN is one of the centers for the development of imaging techniques and electronics. Georges Charpak's multi-wire proportional chamber revolutionized the technology to detect charged particles in physics experiments; it also made new imagining techniques possible. The first positron emission tomography (PET) scanner, the key technology in modern imaging methods, was developed by CERN physicists. Small proton accelerators have been used since the 1990s for tumor treatment. Proton therapy is far more specific and precise in depositing radiation energy than traditional methods. This helps in treating tumors in the eye or in the human brain.

Scientific developments and technology transfer are the main objectives of CERN today. The organization promotes the cooperation of scientists from different nationalities beyond any political orientation. This is the language that is understood all over the world. All of their endeavors serve the great mission of

Aerial view of CERN's main Meyrin site, Switzerland.

CERN to find answers to the great questions of mankind: who are we? Where do we come from? Where are we going?

International Relations of CERN

- The CERN Convention was ratified by 12 countries. On September 29, 1954, the European Organization for Nuclear Research officially came into being. The 12 founding member states are Belgium, Denmark, France, the Federal Republic of Germany, Greece, Italy, the Netherlands, Norway, Sweden, Switzerland, the United Kingdom, and Yugoslavia.
- 1959–61 Austria and Spain join the membership. Yugoslavia leaves the organization in 1961 for financial reasons.
- 1969 Spain leaves the organization and rejoins the membership in 1983.
- 1991 Finland and Poland join the membership.
- 1992–3 Hungary, the Czech Republic, and Slovakia join the membership.
- In 1999, Bulgaria becomes the 20th member state of CERN.
- Member states (2013): Austria, Belgium, Bulgaria, the Czech Republic, Denmark, Finland, France, Germany, Greece, Hungary, Italy, the Netherlands, Norway, Poland, Portugal, the Slovak Republic, Spain, Sweden, Switzerland, and the United Kingdom.
- Candidate for accession: Romania.

The CERN Convention, signed at the sixth session of the CERN Council in Paris from June 29–July 1, 1953.

- Associate member states in the pre-stage to membership: Israel (from September 2011) and Serbia (from December 2011).
- Cyprus became an associate member in 2012.
- Ukraine became an associate member in 2013.
- Observer states: the European Union, India, Japan, the Russian Federation, Turkey, UNESCO, and the US.
- Non-member states involved in CERN programs: Algeria, Argentina, Armenia, Australia, Azerbaijan, Belarus, Brazil, Canada, Chile, China, Colombia, Croatia, Cuba, Egypt, Estonia, Georgia, Iceland, Iran, Ireland, Jordan, Lithuania, Macedonia, Mexico, Montenegro, Morocco, New Zealand, Pakistan, Peru, Saudi Arabia, Slovenia, South Africa, South Korea, the Republic of China (Taiwan), Thailand, the United Arab Emirates, and Vietnam.

The Future of CERN

The LHC was shut down in February 2013 for 18 months to undergo maintenance. The machine will be restarted in early 2015 with its operating energy to be doubled to around 14 TeV of collision energy or 7 TeV per beam. The LHC will be in operation until at least 2030. The installation of more powerful magnets and other equipment will allow for better focusing, thereby considerably increasing luminosity and the number of collisions. At CERN there are already plans to build a Very Large Hadron Collider (VLHC), an even larger ring accelerator. According to the calculations, an accelerator that provides for 40 TeV of collision energy, three times the performance of the LHC, would have to have a circumference of about 200 kilometers or 65 kilometers in diameter.

On the other hand, there seems to be a consensus in the global community of physicists that the next generation of accelerators will be linear accelerators to collide electrons and positrons. The energy of the accelerated particles will be between 0.5 TeV and 1 TeV. Although this is much less energy than in the LHC, a powerful electron-positron accelerator would complement the capabilities of all the other experiments at CERN. For example, detailed research regarding the properties of the Higgs boson—and maybe other particles—would be made possible. There is intense debate over the proposal for the International Linear Collider (ILC) in Geneva. DESY in Hamburg has also developed a model for a new linear accelerator as part of the TESLA Technology Collaboration. The ILC would be between 30–50 kilometers long; the estimated cost of building the ILC is in the region of 7 billion US dollars (www.linearcollider.org).

The History of CERN

- September 19, 1946: Sir Winston Churchill's speech is given at the University of Zurich, expressing the idea that Europe "is the origin of most of the culture, arts, philosophy and science both of ancient and modem times."
- 1949: A group of European particle physicists (Pierre Auger, Niels Bohr, Edoardo Amaldi, Raoul Dautry, Louis de Broglie, and Lew Kowarski, among others) discuss the creation of a common laboratory.
- May 1950: Isidor Rabi, an American physicist, proposes a European initiative to UNESCO.
- Convention of CERN, 1953: It is decided that the European Organization for Nuclear Research will be headquartered in Geneva. "The Organization

Aerial view of CERN, Geneva airport, and the LEP/LHC tunnel. (© CERN, LHC-PHO-1986-001)

shall have no concern with work for military requirements and the results of its experimental and theoretical work shall be published or otherwise made generally available."

- 1954: The 12 founding states allow for a budget of 130 million Swiss francs (for the period 1954–1960).
- 1960: The Proton Synchrotron goes live under Director John Adams. Construction of large bubble chambers (BEBC and Gargamelle) begins.
- 1973: Discovery of weak neutral currents in the Gargamelle bubble chamber, confirming the electroweak force of the Standard Model. Electronic detectors developed by Georges Charpak (multi-wire proportional chamber).
- 1970s: Construction of the Super Proton Synchrotron and the Intersecting Storage Rings.
- 1983: Discovery of the W and Z vector bosons by Carlo Rubbia and Simon van der Meer. The Standard Model established.
- 1983–1989: Construction of the Large Electron-Positron Collider tunnel begins; it is to be 27 kilometers long. It was the largest civil engineering project in Europe at the time.

- 1989: Tim Berners-Lee develops the World Wide Web at CERN.
- 1994: Construction of the LHC starts (budgeted at 6.5 million Swiss francs).
- September 10, 2008: The LHC is commissioned.
- July 4, 2012: The discovery of a "Higgs-like" particle is announced.
- The 2013 Nobel Prize in Physics is jointly awarded to François Englert and Peter W. Higgs "for the theoretical discovery of a mechanism that contributes to our understanding of the origin of mass of subatomic particles, and which recently was confirmed through the discovery of the predicted fundamental particle, by the ATLAS and CMS experiments at CERN's Large Hadron Collider."

2 The Practitioner: Rolf-Dieter Heuer

Rolf-Dieter Heuer,
Director-General of CERN.
(© Michael Krause)

Rolf-Dieter Heuer was born on May 24, 1948 in Boll (now Bad Boll), Germany. He studied physics at the University of Stuttgart and received his PhD in 1977 at the University of Heidelberg. Dr. Heuer is an experimental physicist. Most of his scientific papers deal with electron-positron reactions, the development of experimental techniques, and the construction and operation of large detector systems. In 1977, he became a research associate at the University of Heidelberg with the JADE experiment. Until 1983, he worked at the electron-positron storage ring PETRA at the German Electron Synchrotron DESY in Hamburg. From 1984 to 1998, Dr. Heuer worked at CERN in the framework of the OPAL experiment at the Large Electron-Positron collider, the LEP. He was coordinator during the construction and start-up of the 1989–92 LEP1 and was the OPAL spokesperson from 1994 to 1998. In 1998, Rolf-Dieter Heuer was appointed to a professorship at the University of Hamburg. In December 2004, Professor Heuer became research director for particle and astroparticle physics at DESY in Hamburg. Since January 1, 2009, Professor Heuer has been the Director-General of CERN.

Question: You were born in Bad Boll. What kind of memories do you have of your hometown?

Heuer: I remember that my parents moved away from there when I was about three years old. I do not remember much from the period before—perhaps visits to relatives, visits to Boll or the Swabian Alb, or going somewhere to the countryside. It was great to visit relatives, to visit a village, to enjoy homemade baked goods, to enjoy the homemade apple juice. But I think I've got a very selective memory.

Question: That was a more rural setting?

Heuer: I am a countryside boy.

"Boller Badt," engraving by Matthäus Merian, dated 1643, one year after the death of Galileo and Isaac Newton's birth (*Topographia Sueviae*, 1643/1656).

Question: How did you come to physics?

Heuer: I do not know. But I was always interested in questions like: what is the world made of ? What are the smallest components of the world? I was always fascinated by the smallest things, and—at least at that time—not the biggest things. For me, it was not the stars in the universe, but just the smallest things. Therefore, I held a lecture at high school about "atomos"; the indivisible, about Democritus' atomic physics, i.e., about the smallest things. I liked to play very close to rivers, brooks or ponds. Even back then I liked to look at the small things.

The Atomic Theory of Democritus of Abdera

The Greek philosopher Democritus (born around 460 BC in Abdera, Thrace, an Ionian colony in Asia Minor; died 400 or 380 BC) was a student of Leucippus. He lived and taught in his native city of Abdera. Democritus is one of the pre-Socratic philosophers (Socrates, 470–399 BC) and is regarded as the last great natural philosopher. Democritus used his family's wealth for extensive travel and research of all kinds. He is boasted as one of the most educated men of his time. The list of his numerous writings demonstrates the full extent of his interests and knowledge. Unfortunately, only fragments of these original writings have survived. Like his teacher Leucippus, Democritus postulated that nature is composed of small, indivisible units: atoms (from the Greek "atomoi"; "atomos" = indivisible). According to Democritus' theory, these atoms were suspended in empty space, or vacuum. Democritus' central message, according to a document by the Greek Galen, a physician and naturalist who lived during the 2nd century AD, was:

> The objects of sense are supposed to be real and it is customary to regard them as such, but in truth they are not. Only the atoms and the void are real.

Democritus' Atomic Model

- All matter is composed of indivisible ("atomos") bodies: they are materially identical, but differ in shape, position, and arrangement. Their possible combinations are theoretically infinite.
- Atoms are solid and indivisible. They have the properties of matter, which is composed of them. Smooth objects consist of round atoms, while rough objects are made of angular atoms.
- Atoms change their movement through mutual touch, pressure, and shock. These collisions can deform the atom, but it quickly returns to its original shape. Between the atoms there is only empty space (from the Latin: "vacuum").
- Atoms react with each other when they interlock or "hook" together; then they appear as water, fire, plants, or humans.

Democritus was called the "laughing philosopher" by his contemporaries. He believed that the human soul should capture the essence of things and thereby attain a cheerful, serene mood. Democritus called this serene, equanimous mood, "Euthymia," the highest good of man.

Democritus' atomic model and his notions of space and matter were vehemently rejected by Socrates (who believed that "the only true wisdom is in knowing you know nothing"), Plato (428–348 BC), and Aristotle (384–322 BC). Democritus' theory of atoms and the vacuum was replaced by the "horror vacui," which was postulated by Aristotle in his work *Physics*, where he examined the nature of matter, space, time, and motion. Aristotle argued that there was a natural aversion against absolute emptiness—empty meaning a place "where nothing exists." Aristotle argued (*Theology*, Book XII) for a force that causes all movement in the world. According to Aristotle, this primal force, the "Unmoved Mover," permeates all space and all time. If this eternal, primordial power exists everywhere, there cannot be "nothing." For Aristotle, it was clear that the heavenly bodies are eternal and imperishable, while the material world is transient and consists only of the four elements: fire, water, earth, and air. The Aristotelian theory of nature, its idea concerning the structure of matter and the universe and its division into two spheres, the earth and the heavens, ruled the ancient world until the emergence of modern science.

Democritus, the laughing philosopher.

Question: Did you discover nature on your own, or how did it happen?

Heuer: I did not discover nature, I've only played. I was seven or eight years old, and it had nothing to do with the discovery of things. That came later, in high school. But I could always play together either alone or in groups.

Question: Your teacher is supposed to have said, "If you're not going to be a physicist, then I don't know anything."

Heuer: No, no this is wrong; he did not use those words. But I was interested in physics and how things work. I was somehow not quite sure which way I should take. More towards physics or more towards mathematics? The curriculum of my physics teacher was based more on logic and the understanding of coherences, more on understanding the logic than just only to learn formulas. This is something that I have learned at that time: you have to know where the formulas are written down, so you can find them. And you have to understand the logic with which the formulas are built—this is what I always try to teach my students.

Question: How did your career develop?

Heuer: I never planned a career. It just went on and on, step-by-step. But first, I had to go into the army, because at that time you had to join the army after high school. One and a half years in the army; that really was a difficult time. Then, I went to university and, frankly, in the beginning I did not understand much there. At the university, there was a completely different way of dealing with things, of how to learn things and find out something. In addition, I did not start in the first semester but in the second. At the University of Stuttgart, at that time, you could only start studying once a year, and not twice as is common today. Another thing that I learned then is this: if you have problems and difficulties, you just have to look around. Then you will see all the others who have the same difficulties. Difficulties are there to be overcome. That is what I've learned, among other things, at that time.

Question: How did you and your colleagues think about it? Do you still know your former colleagues, are you still friends with them today?

Heuer: I realized that we are all in the same boat. That helps and motivates. And with some I am still friends. Some of those with whom I formed small working groups to study.

Question: If you look back, what were your highlights at that time, the outstanding events during your studies?

Heuer: The fact that I was finally able to understand a few things...[laughs]...I think that was really a highlight. Really, I finally understood that I could do that. I could understand things that I could explain to others in a reasonable time—and then take the next step.

Question: What do people do at CERN?

Heuer: Some do basic research. But, on the other hand, you cannot conduct any basic research if you do not innovate in the technological field. You also need tools, so-called "apparati," to work with. Others build the buildings, while others maintain the equipment, like the accelerator, etc. Still others take care of the infrastructure; we do nothing else here than in ordinary, everyday life. Infrastructure, road maintenance, street cleaning...we have all sorts of people who are doing everything to ensure that scientists can conduct their research. But the basic mission is of course to increase knowledge, to bring humanity more knowledge and hopefully thereby improve humanity. And we do that, a unique thing here at CERN, within a totally global environment. We have registered almost 100 nationalities as scientific staff here and they all work together very well. They all have the same motivation, to create more knowledge through research.

Question: Is there such a thing as a common spirit at CERN?

Heuer: Absolutely. A part of the CERN spirit is that you can work together regardless of nationalities and regardless of the culture or different cultures. "Regardless," this is the important word here. If you leave out politics, then it works.

Question: Is CERN something like a role model to solve problems elsewhere in the world?

Heuer: I think this is possible and I think that CERN already does that. CERN was founded in 1954, but the first discussions started as early as in 1949, just a few years after the end of the Second World War. At that time, people began to talk about a European lab. Representatives of different nations sat together at one table—nations which had previously been at war against each other. This is proof that such a thing is possible and that we can make progress. Perhaps sometimes there are only small steps towards peace or understanding among nations. CERN has done this during its entire history. When the Iron Curtain still existed people from both sides have worked together here at CERN. This worked very well, and today we have people here from Pakistan working together with people from India, China, or from both sides. All this shows that we can work together peacefully. I do not know if you could call it a role model, but at least it is a small step on the way to a better understanding of people and the world.

Question: We have been talking about the positive things so far. How do you deal with obstacles?

Heuer: That depends on how you define the word "obstacles." Some obstacle is of human nature. If you have two and a half thousand employees, and ten thousand scientific users, then you are like a large village with many, many visitors. There are always minor interpersonal problems; that's completely normal. And so I deal with it, normal. Basically, obstacles do not turn up necessarily between people from different cultures. They exist mostly between people from the same culture, you have to know. But these are the normal things that you also face at every university, in any company or institution. Then, of course, there are technical problems; sometimes things just do not work as originally planned. But there are very good people here to solve these problems. It is not me who determines everything, because I am probably not an expert in the matter. I need to find out who the expert is. I need to talk to the expert and trust him; this works fine. So leave it to them to solve technical problems. And then there are financial problems from time to time, of course, especially in times like this. The main thing is not to be too upset. You deal with these problems calmly; this method seems to work.

Question: There are new movements in the world, such as "Occupy" for example. Can CERN help to save the world?

Heuer: This is an ambitious goal. We can supply only a small component of the whole building, of the world of tomorrow, so to say. This module is to develop technologies that will allow us to conduct research. Without research, we would not have the world as it is today. We would not meet today's standard of living. People should never forget that what we have today is the result of applied sciences and of basic research, which has been done several decades ago. And we must do the same for future generations. The other building block that we can provide is an understanding among very different people. In most cases it works like this: first you have colleagues, then they will become your friends. This will certainly lead to a better understanding of each other. It is about two things: basic research for the more distant future and developing technologies for the near future—and the better understanding among people.

Question: What is your role at CERN?

Heuer: In the first place I consider myself as a lubricant of the mighty wheels of CERN, so hopefully things go smoothly. And at the same time I'm also the flywheel that drives the wheels—with the help of everyone, of course. A management without the people doing the real work, this would be nothing. If people do their work without the help and understanding of the management that is not good either. I think that all of us, together, we are the wheels of CERN. I see myself as *primus inter pares*.

Question: You have known CERN for many decades now. What changes have occurred for you?

Heuer: When I first worked at CERN, I supervised one of the large experiments. I did not have too much contact with top management because everything went smoothly. Only if something did not work as it should have, then you had to get in contact with the top management. I returned to CERN in a completely different position. That was quite strange in the beginning. The people with whom I had worked before, they did not know how they should deal with the new situation. Are you still the same? Have you changed? But it turned out to be a fast mutual learning process for both sides, and then everything worked fine again. My main job today is to look after the people and to keep them away from problems so that they can do their research. If they don't have any problems, then I have done my job well.

Question: What are the biggest problems here?

Heuer: On the one hand, the biggest problem is to make sure that people work together well, of course. If you involve people at a very early stage, then they will take responsibility when necessary. That might be against their intentions, but they will take responsibility for the common good. To deal with people is, I would not say, the biggest problem; but it makes the most trouble if you are responsible for the functioning of an institution. That's the main thing because all other things are of a technical nature and you can always solve technical problems. As long as there is logic and technology involved, fine. If feelings are involved, it becomes difficult.

Question: You are a fan of VfB Stuttgart. Do you have more hobbies?

Heuer: I do not particularly like the word "fan," but yes, my favorite is still the club VfB Stuttgart. Although times have changed, of course. In this year's season they are not as bad as in the one before and that is a good thing. I have many interests. But, in principle, one could say that I have no time for them. You could say that I am interested in sports, but I don't do it anymore. I lie on the couch and watch or listen to sports programs. This is my sports and it is relaxing indeed. I also like to travel. I would very much prefer to travel for my own pleasure, but at the moment I have to do it because of my job. But since I love traveling this is not so bad. I also love hiking. My wife would say now, "No, no, no, I do not believe you." Because sometimes it is rather difficult to prompt me to get up and go. And there are those lower instincts that often defeat the urge for mobility [laughs]. But when I'm out there again, I love to hike in the Alps or something. And I have my model railway; I just like the small things.

Question: Within the history of science, where would the LHC be?

Heuer: I think that this could be the machine that allows a view of the dark universe for the first time. Because what we see around us is just about 4–5% of the mass and energy density of the universe. Ninety-five percent are unknown to us, so "dark" is in quotation marks. A quarter of this dark universe is dark matter, and the LHC could actually bring the first light into this dark universe. That would really change our view of the early universe, when visible and dark matter together did shape our universe. This also depends on whether we will find the famous Higgs boson or not. That will be a milestone in our understanding of the microcosm and the early universe as well. So I think the LHC is a very, very important tool, but to put this all together into one picture is a very, very difficult task. One hundred years ago, we saw very small experiments that were fundamental—not in a technological sense, but in a scientific one. In this sense I would rank the LHC very high. But please, ask me again fifty years from now.

Question: Ninety-five percent is a lot indeed. If we know nothing about it, then we do not know about a whole lot.

Heuer: Right. And there is probably more that we do not know of. Now it gets philosophical, really. We know that we do not know about 95% of everything. And what we also do not know, we do not know of. It might be much more that we know nothing about. And much more around it, of which we also do not know anything. But this is what fascinates all of us—at least all of us physicists. We had

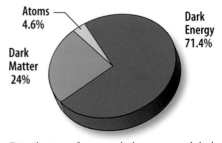

Distribution of matter, dark matter, and dark energy in the universe. (© NASA/WMAP Science team, 2008)

to work for three to four decades, so that we could explain 4–5% of the energy density of the universe. If we would really enter the dark universe with the LHC, it might not be a quantum leap exactly but it would be a giant leap forward. I hope that we will achieve this within a decade or so.

Faust

No dog could live thus any more!
So I have turned to magic lore,
To see if through the spirit's power and speech
Perchance full many a secret I may reach,
So that no more with bitter sweat
I need to talk of what I don't know yet,
So that I may perceive whatever holds
The world together in its inmost folds,

See all its seeds, its working power,
And cease word-threshing from this hour. [...]
Into the whole how all things blend,
Each in the other working, living!
How heavenly powers ascend, descend,
Each unto each the golden vessels giving!
On pinions fragrant blessings bringing,
From Heaven through Earth all onward winging,
Through all the All harmonious ringing!

(*Faust* by Johann Wolfgang von Goethe, translated by George Madison Priest)

Rembrandt Harmensz van Rijn (1606–1669): *A Scholar in His Study ('Faust')*, about 1652, Rijksmuseum Amsterdam.

Question: How could the LHC facilitate a peek into dark matter territory?

Heuer: Dark matter must be some sort of particle which interacts with regular particles only very weakly. Within the energy range of the LHC, there could be particles that we can create thanks to $E=mc^2$, Einstein's formula—sorry! "Creation" is the wrong word; someone else already has this job! "Produce," I mean, we can produce. We could produce dark matter particles or new particles, which then could turn out to be dark matter particles at the end.

Question: Is the Higgs such a particle?

Heuer: No, not the Higgs. The Higgs is still part of the so-called Standard Model, which we need for the 5% we just talked about. Actually, we do not understand that 5% completely. But we have examined the so-called Standard Model for sure. This model explains that 5%, and it passed all tests pretty well. If the Higgs really exists, then we have completed the Standard Model in some sense. If we discover that it does not exist, then our theory about that 5% gets a big hole. One of the cornerstones of the theory would collapse. Then it would be the task of the LHC to find a replacement for this cornerstone. And there is much more to discover.

Simulation of a collision. (© 2000 CERN)

Question: Is this "new physics," physics beyond the Standard Model?

Heuer: New physics is the physics that goes beyond what is valid now. If we, for example, discover the Higgs particle, then that is no new physics, because it is part

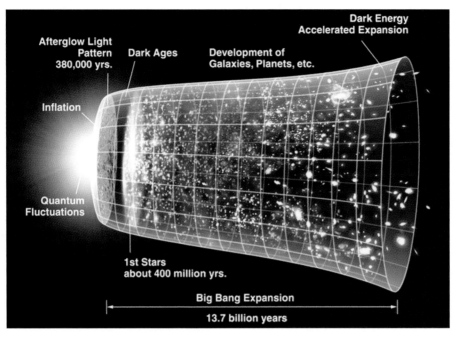

The universe according to the Standard Model. (© NASA/WMAP Science Team)

of the Standard Model. For me, "new physics" is anything beyond the Standard Model so there must be another model that needs to be built around the Standard Model, which allows approximations into lower energy regions.

Question: Is there a world outside of our world?

Heuer: Everything we know is within our world. If you were to find extra-dimensions, that would still be within our world. It does not help much to imagine something. Our imagination does not go beyond three dimensions, for instance. It is very hard to think about reality and imagination beyond. Sometimes it's hard to even imagine what is behind reality [laughs]. If you define an elementary particle as a point-like particle—which in itself would be wrong—that has mass; that is, it is a massive entity and it has a spin. How could you imagine a point-like elementary particle? I cannot imagine that! But I can put it mathematically, thus describing the physics of a microcosm. It is one thing to describe and formulate how everything fits together. For our imagination this is rather difficult. Some people think that they can very well imagine multiple dimensions. They look at the stars and actually they cannot imagine what it is. The same is true for small dimensions—it's just very difficult to imagine such a thing. Maybe future generations can.

Question: There is a nucleus and an electron, which is very far removed from the core. What's in between?

Heuer: It is the old model that the electrons orbit around the nucleus. Actually, there is rather a probability of the presence of the electron; there are forces. If you want to look at the matter, then we are all mostly empty. It is the forces that make up the mass and they make us, too. But is it easy to imagine a force field? You can describe it though, but can you imagine it?

Question: Can you imagine the atom and its structures?

Heuer: You can describe this with relatively simple formulas—but to put it in words?! Perhaps this is also relatively easy, but if you try to imagine it right before your eyes then it starts getting difficult.

Question: Goethe wrote, "So that I may perceive whatever holds the world together in its inmost folds." You are all looking for that, right?

Heuer: I intend to see the forces that act between elementary particles. So, for me, what holds the world together in its inmost folds—it is the forces that act between elementary particles. If something does exists around it—something that really holds the world together, then I do not know it and I will not engage in this question. Because then it is no longer science or research, then it becomes will and imagination, which is an interesting field certainly.

Question: The research at CERN recreates the Big Bang...

Heuer: We will never recreate the Big Bang; we just get very close to it. We study the evolution of the early universe and not the Big Bang. Although, the LHC could be called a kind of "big bang machine," the thing is that shortly after the Big Bang, everything happened that has determined our universe. That is why it is so interesting to come close to the Big Bang and we get close to a millionth of a millionth of a second, i.e., 10^{-12} seconds. On a human scale this is extremely short. But if you look at the evolution of the universe, this is a very long period of time because very much happened. The closer we get to the Big Bang, the easier the extrapolations and the more accurate measurements we can take. The LHC is nothing more than a super-microscope in order to see the smallest objects. If you look at these smallest objects, then you get as close as possible to the early universe.

Question: Are you satisfied with your work here?

Heuer: I think that you can already be happy if you bring something to an end or if you can find an answer to one of the many questions in physics, which in turn will generate new questions. Then you can be satisfied. But that could also be dangerous because when you are satisfied, you will stop. I think you should never be completely satisfied. There must always be something left, so you can continue. But if you can successfully deliver something that you wanted to deliver, then you can feel a little bit happy.

Question: What motivates you?

Heuer: What motivates me is that I really want to find out a little bit more about the early universe and the microcosm. If humanity should stop to want to know more about the basics of our lives, then we cease to be human. People have this urge, and this is what drives me also.

Question: What would you choose as a symbol of the universe? [Onion, walnut, or pomegranate?]

Heuer: As a symbol of the universe, probably the pomegranate. It contains many items that look the same, but they are all a bit different. This shows the variety of the universe, perhaps. There are many galaxies there. For me, the walnut is too... the structure inside is very complicated, but once cracked ... and the onion is just too simple. OK, that would be the pomegranate for me.

Rolf-Dieter Heuer with a pomegranate. (© Michael Krause)

Notable Quotes by Rolf-Dieter Heuer

- Without research we would not have the world as it is today.
- Our basic mission is to expand human knowledge and hopefully thereby improve humanity a little bit.
- What holds the world together are the forces that act between elementary particles.
- If humanity should cease to want to know more about the basics of our lives, then we cease to be human altogether.
- We know that we do not know about 95% of the universe. And we also do not know what we do not know. Out there, there may be many more things we know nothing about and there might be much more around it, of which we know nothing, too.

3 The Beginning of Modern Physics: Galileo, Copernicus, and Kepler

Galileo Galilei (1564–1642) came from an impoverished Florentine patrician family. His father, Vincenzo Galilei, was a musician and music theorist. He introduced young Galileo to the mysteries of mathematics but told him to choose a more lucrative career as a doctor. In 1580, Galileo began studying medicine at the University of Pisa. Four years later he moved to Florence to study mathematics. In 1589, he returned to Pisa to become a lecturer in mathematics. Galileo supplemented his poorly-paid job by selling self-made instruments, such as simple yet accurate thermometers. In 1592, Galileo Galilei was appointed professor at the University of Padua. He began to study the pendulum laws and the laws of acceleration. Galileo's experiments at the Leaning Tower of Pisa are an unproven legend. Some particle physicists, however, refer to Galileo's experiments in Pisa as "the first particle accelerator in the world" (Leon Lederman).

Galileo Galilei is often called the "father of modern science." He made fundamental contributions to the scientific revolution of the early modern period. His experiments and instruments (apparati) were ground-breaking, both in fundamental as well as in applied sciences. Galileo basically criticized the purely theoretical approach of the doctrine. At that time science, i.e., physics, was only taught theoretically, not through experiments. Galileo, however, wanted to observe and measure phenomena in nature, independently and critically. Therefore he conducted experiments and therefore he needed instruments. Numerous appliances and scientific instruments were developed between 1550 and 1700. They were designed to serve the exact nature of the investigation, regardless of the scientist's imagination. Galileo interpreted the experiments he had carried out with the aid of mathematical formulas. He wrote his most important work ("Dialogo," 1632) not in Latin but in plain Italian so that his teachings could be understood universally. Galileo was a scientific revolutionary at the center of a new era, the spiritual and cultural awakening of the Renaissance and the rediscovery of the ancient world. With his approach, Galileo radically changed the hitherto valid Ptolemaic world view.

In 1609, Galileo manufactured special lenses in his workshop; they were made of Murano glass. He inserted the lenses into a long, wooden cylinder, like the one that the Dutchman Jan Lippershey, the inventor of the telescope (from Greek, meaning: to look further), had created. With his improved telescope Galileo discovered the four largest moons of Jupiter, the composition of the Milky Way from individual stars, and sunspots. His observations ultimately led to a revolutionary conclusion: the Earth is not at the center of our Solar System, the Sun is. Thus Galileo Galilei gave proof to support the heliocentric system experimentally (by observation) and with the help of an apparatus (telescope). From then on, the Sun was at the center of our world—in contrast to the previously valid, 1500-year old Ptolemaic world view in which the Earth was at the center of the universe.

Galileo's observations had been picked up by the German physician, mathematician, and astronomer Nicolaus Copernicus (1473–1543) in his work "De Revolutionibus Orbium Coelestium." Copernicus conducted his astronomical observations in his spare time as a "hobby." His observations led Copernicus to hypothesize (mathematically incorrect) circular orbits for the planets. In his system, the Earth rotated around its own axis once a day and all of the other planets revolved around the central Sun. The heliocentric world view, and a spinning Earth, however, totally contradicted the doctrine of the Catholic Church. In 1539, Copernicus published his work. Little is known about the distribution of Copernicus' "De Revolutionibus" and the so-called "Commentariolus," an earlier, non-argumentative summary. What we know is that Tycho Brahe, a Danish astronomer and native from a noble family, received a copy of "Commentariolus" during the coronation of Emperor Rudolf II in 1575.

In 1599, Tycho Brahe (1546–1601) became a mathematician and astronomer at the Prague court of Emperor Rudolf II (1552–1612), a major patron of arts and sciences. Tycho's assistant and successor was the young German astronomer Johannes Kepler (1571–1630). Kepler had drawn great scientific attention with his 1596 publication, "Mysterium Cosmographicum." This great book dealt with the "secrets of the universe," and it was dedicated to the stylish and brilliant idea of world harmonies based on the geometric structures of planetary orbits from Platonic solids (cube, tetrahedron and octahedron). Brahe, nicknamed "the man without a nose," had in turn developed his own world-system. It was a mixture of a heliocentric and a geocentric model. Brahe, who had made his observations without the benefit of a telescope, died before completion of the Prague observatory.

In 1609, when Galileo Galilei pointed his telescope to the night sky and the stars, Kepler published "Astronomia Nova," subtitled "Physica coelestis"—i.e., "physics of the heavens." Kepler discovered that the planets did not move on circular tracks, but had elliptical paths instead. All of the tracks had one of their foci in the Sun. Now the mathematically calculated orbits were in line with the real observations. The end of the old Ptolemaic system had definitely arrived, even if anything else was completely contrary to the dogma of the almighty Catholic Church.

Johannes Kepler's mathematical formulas for planetary movements are still valid today. They apply not only to planets but to all objects that move in gravitational fields. Any missile in space obeys Kepler's laws. Kepler himself did not see his pioneering work as scientific laws but as hypothetical formulas that served to further explicate the great harmony of divine God. Kepler went on to work on esoteric research in order to unite his physics with the Pythagorean theory of harmony. His main work, "Harmonices Mundi" ("The Harmony of the World") was published in 1619. In this treatise, Kepler, a devout Catholic, tried to prove that the universe is part of a divine harmony.

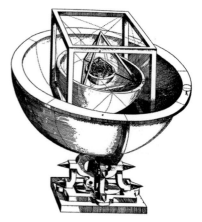

Kepler's model of the Solar System, "Mysterium Cosmographicum" (1596).

Galileo Galilei chose another way. He totally opposed the Church with his book, *Dialogo di Galileo Galilei sopra i due Massimi Sistemi Del Mondo Tolemaico e Copernicano* (*Dialogue of Galileo Galilei on the two major world-systems, Ptolemaic and Copernican*), published in 1632. In this book, Galileo challenged the almighty Catholic Church quite clearly by proving the correctness of the heliocentric, Copernican system beyond any doubt. The Catholic Church had expressly warned Galileo about the publication. Consequently, the church defended her dogma and arrested Galileo. Although he retracted his statement one year later, he had to spend the rest of his life under house arrest. Galileo's proof for the heliocentric system was consistent, however, and therefore final. The ancient Ptolemaic system with the Earth at the center of the world was done away with by the work of Copernicus, Kepler, and Galileo.

4 The Experimentalist: Tejinder S. Virdee

Tejinder Singh Virdee.
(© Michael Krause)

Tejinder Virdee was born in 1952 in Nyeri, Kenya. In 1967, his family, Indian by origin, moved to the UK. Tejinder Singh (Jim) Virdee studied physics at the Queen Mary University of London. In 1974 he went on to complete his PhD at Imperial College London while working on an experiment conducted at the Stanford Linear Accelerator Center in the US. Dr. Virdee joined CERN as a postgraduate in 1979 as a member of the UA1 collaboration at CERN's proton-antiproton collider (SPS). From the early 1990s, Dr. Virdee developed the concept and technology for the CMS experiment and is often referred to as the chief architect of the CMS, "a giant 3-D 100 megapixel digital camera weighing 14,000 tons, which takes 40 million pictures per second" (Virdee). From 2006–2010, he led the experiment during the critical phase of its final assembly and during its first tests and trial runs. The CMS experiment is an all-round machine that is used to search for the Higgs boson and dark matter. More than 3,000 participants from 38 countries work within the CMS framework. Dr. Virdee's work currently focuses on adapting the CMS collaboration for the LHC's future energy levels.

Since 1996, Dr. Virdee has been a professor at Imperial College London. In 2009, he received the James Chadwick medal and since 2012 Professor Virdee has been a Fellow of the Royal Society. For services to science, Tejinder Virdee was honored Knight Bachelor by Queen Elizabeth II on June 14, 2014.

Question: When you looked at the sky as a child, what did you see? What questions did you ask yourself?

Virdee: Well, I was born in Kenya, so my recollection from that time is that the skies used to be very clear and bright, and you could see the Milky Way. It was on the equator as well, so I was fascinated about what was going on in the natural world because it's also a beautiful world in terms of the scenery as well. I think one asked these questions and of course one doesn't know the answers, and then one continues one's studies and so on and so on. So eventually my family moved to England and there I started doing science and this answered individual subjects. I had a very, very brilliant physics teacher who really sparked me to thinking about physics as a subject to study, because it gives you a feeling of how nature works on a fundamental level, so that's how it all started in a sense.

Question: It was a personal relationship to your teacher that encouraged you to study physics. How was that relationship?

Virdee: It was fascinating, because we had just moved from Kenya to England. I still know his name; it was Mr. Stockley. He made us feel very much at home because we had come to a new country and he was a fascinating teacher. He knew his physics; he didn't need any notes. He just asked the class: what was the last sentence I spoke that you wrote down? And from there on he would run the lesson. But he was also a very fascinating cultural person. He used to take students to Italy every Easter for visiting Florence, Rome and Taranto—that's what I did—and there you'd visit the art galleries, also the Vatican, also all sorts of interesting places from a few thousand years ago and so on. It was fascinating. So it was pretty much a rounded education I was getting with this physics teacher.

Question: After school, what came next?

Virdee: I think what I felt was that I wanted to do physics and so I went to the University of London to do my first degree, and after that I got fascinated by particle physics, because that was the field I felt that was the closest to trying to understand how nature works on the fundamental level that I mentioned earlier.

Question: What is that fundamental level in particle physics?

Virdee: In fact, "particle physics" is a modern name of a centuries old effort to understand the basic laws of nature. So this quest was started let's say at Newton's time or Galileo's. When Newton combined celestial and terrestrial gravity, [...]

that was one of the first intuitions that phenomena that we would see here on Earth could actually have some relation with other phenomena—and so we have this quest for unification. One of the primary goals of physics is to understand the wonderful variety that we see in physical phenomena or natural phenomena in a unified way.

In the mid-1800s the next big step took place, which was the understanding of electricity and magnetism, and combining this with the theory of optics, for example. And from there we have essentially the Industrial Revolution because electromagnetism played a key part in the Industrial Revolution. And later on in the century J.J. Thomson discovered the electron. The discovery of the electron is the basis of electronics, if you wish. The word "electronics" has an "electro" in its spelling. It's the mastery of the movement of electrons, which were first discovered in a sense only 115 years ago. And how can we live without electronics?

So there are very many instances of this nature—quantum mechanics, relativity, lasers, for example—all these things come from fundamental science. So in a sense, when one looks at the development, we don't live in the same way people lived at Newton's time, for example. And the way we have actually altered our way of living has been through steps in our understanding of nature. Fundamental science is actually at the heart of human development, in a sense.

Question: In the 20th century, what were the next steps?

Virdee: In the last century we had the advent of quantum mechanics, which is essentially the understanding of chemistry in atomic physics. Then we had the revolution of space-time and gravitation which Einstein developed. And then we had powerful machines called accelerators coming up, and they could allow us to see deeper into matter. And when we looked deeper into matter we found that the atoms had not only electrons and nuclei but the nuclei had neutrons and protons, and the neutrons and protons themselves have constituents called "quarks" and "gluons."

So we have been able to look deeper into matter, and by looking at that we found very large numbers of particles. And in fact, today, we know that we can actually describe all of these things with a small set of particles called quarks, gluons, and the bosons, which mediate the forces. So this is a sort of new periodic table, if you wish, and it is much smaller than the periodic table of chemical elements. This now is within the Standard Model, which contains the constituents of matter, and also the forces that control their behavior. In that sense the culmination of one of the key features, or key outcomes of 20th-century science, has been the construction of the Standard Model.

When I started as a graduate student, half of the particles in the Standard Model had not been observed so a lot of progress has been made since we've started designing the LHC experiments: the discovery of the top quark, the discovery that the neutrinos have a small mass, and so we are waiting for the next big step. In essence what we are trying to do is to understand nature in a unified way. What does that mean? All of the physical phenomena that we observe in nature are governed by essentially four forces. Two are very familiar to us, which is gravity and electromagnetism. The third one is the weak attraction which powers the Sun. So what's happening inside of the Sun is determined by weak attraction and the interaction that holds the quarks and gluons inside the proton or the neutron, called the "strong interaction."

Now, we have quantum theories of three of the four forces. Electricity/magnetism and weak attraction we consider today to be unified. What it means is that we have been able to go to such high energies of high temperatures—go back in time in a sense—and at those temperatures or energies we can distinguish between a process that occurs through weak attraction or through an electromagnetic interaction. In fact the probabilities become roughly the same and it is in this sense that we say that it is unified. Now, the strong interaction, although we have a quantum theory of gravity, how does it actually fit into the fold? We have ideas, but these may only work once gravity is brought into the fold. Gravity has been very difficult to bring into fold; still today we don't have a quantum theory of gravity. So what we are trying to do at the LHC is to try to find the next steps towards this goal. In fact, we're again still on this path that started with Newton several centuries ago, and we haven't finished.

The Four Fundamental Forces (Interactions)

- The strong nuclear force (interaction) binds atomic nuclei together. This fundamental force binds the quarks together in hadrons (particles composed of quarks). This force is extremely energetic, but very short-ranged. The carriers of this strongest of all elementary forces are the gluons.
- The weak nuclear force (interaction) is 10^{-13} times weaker than the strong nuclear force. This fundamental force is responsible for beta decay (radioactivity) and is crucial for the fusion of two hydrogen atoms to form helium within the process

that powers stars like our Sun. Carriers of the electroweak force are the W and Z bosons (see Chapter 13 "The Nobel Prize Laureate: Carlo Rubbia").

- The electromagnetic force acts between electrically charged particles. Electrons are held in their orbits around the nucleus by an electromagnetic wave mechanism. This mechanism determines the chemical behavior of matter. The electromagnetic force is responsible for almost all phenomena of everyday life. Carriers of this force are the massless photons.
- The gravitational force (gravity) is responsible for the mutual attraction of masses. This fundamental force acts always and everywhere. Because of it, the apple falls from the tree. Gravity is the weakest of all elemental forces. As of today, gravity cannot be combined with the other three forces; there is no unified theory. The carrier of gravity is the (hypothetical) graviton.

Question: Columbus went to find the sea route to India, and he landed in the Americas. What about the LHC? Where did it begin, where is it now?

Virdee: In a sense, when you look at what Columbus and his colleagues did, the first thing they had to do was to have the vision that there was something out there. And so we believe there is something out there which we need to find. And this will allow us to take the steps forward in understanding how nature works.

The second thing they needed was very reliable and robust ships, and also very good navigators. That analogy would be the accelerator, and the experiments are essentially the ship. So what has happened since we started collecting data—colliding protons about a year or so ago—was that we landed on the shores of America, let's say, if you want to use that analogy, and now we've started exploring and we have probably gone about a hundred kilometers inland, and we're trying to see what the territory looks like. So far it looks like what we have expected but there is a whole continent which has still to be discovered. We don't know yet whether there is a Grand Canyon out there which would be spectacular. So that's where we are at the moment. This process that we started about twenty years ago, in conceiving, designing, and constructing the machine, the detectors, we have another 20 years to go to experience the full exploration if you wish. So we're halfway in this quest, this voyage of discovery.

Question: What keeps you going? Where does your energy come from?

Virdee: There is a sense of accomplishment. These instruments are very advanced instruments; they are the most complex scientific instruments ever built. And we

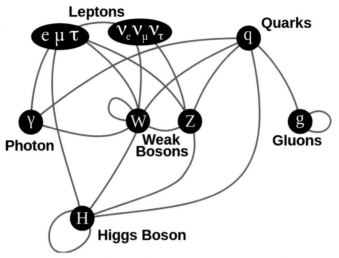

Elementary particles and their interactions. (© Euan Richard)

expect something to happen. We expect to make discoveries so this expectation is what keeps you going—because if you firmly believe there is something out there, which will tell you how nature works and how to take the next step.

Question: Does this make you content?

Virdee: I would rather say "expectant." You anticipate and then … the problem with nature is that—not a problem, just the way nature works—is if you want to find its secrets you have to work very hard and very diligently and very correctly. And so that takes time. But we believe that we will find some of the secrets of nature, which as I mentioned earlier, will tell us what next to do.

Question: Is there a place for religion then?

Virdee: I tend not to discuss those things.

Shiva (Sanskrit for "The Auspicious One") is one of the most important gods of Hinduism. Shiva is considered the most important manifestation of the Supreme. Shiva is often represented as Nataraja, the King of Dance. He dances on apasmâra, the "demon of ignorance." Shiva destroys ignorance—and the universe, thus recreating it. Shiva has four, eight, or more arms that represent his diverse activities. Shiva's dance symbolizes the source of all movement within the cosmos. His dance frees people's souls from their illusions.

Statue of Shiva, CERN, Building 40 (ATLAS/CMS-building). (© Michael Krause)

Question: You once said that only international collaboration makes experiments here at CERN possible. Why is that so?

Virdee: I think this is true. I think no single country or a single region could have made the experiments and the accelerator that we have. Because these are cutting-edge scientific instruments, there were lots of risks; many of the technologies had to be pushed to the limits and some of the technologies didn't even exist at the time. So this was a very ambitious project. For example, in CMS, or in ATLAS, each one has about 3,000 scientists and engineers from about 40 countries. The detectors have been built, or parts of them have been built in the countries. There was also a massive logistic problem, if you wish. But I think some of the satisfying things is to see that people from different countries and regions, different cultures, can really work together and produce an object which is performing wonderfully well, according to the expectations that we had even 20 years ago, and that is extremely satisfying to see. And it has all sorts of repercussions on the line, so in that sense I think it has been necessary and it has been very worthwhile.

Question: People working here are kind of building a society in itself, a technological society. Could that be a role model for the world?

Virdee: I think one has to look at what sort of problems a model like CERN could work. The goal here was to try to understand nature and so you build an instrument and everybody is actually quite keen to do the right thing: to make sure that the instrument will work correctly, as we designed it. So people took great care to make sure everything was done properly.

If there were problems, they were identified, and declared, and resolved collectively. So, in a sense, there has been a goal and that has been concentrating the efforts of these people. That's one thing to remember. There has to be some sort of a common goal which is quite important when many people come together.

Now, there are problems which can lend themselves to models like CERN. I am just sort of thinking aloud, for example, big problems, big issues, big discussions of climate change. Perhaps something like that could be developed at an institution like CERN where the scientists are working on climate change can come together and the goal is clearly to understand what's happening with the climate and then be very dispassionate about understanding the causes and the effects and so on, which can lend themselves perhaps to experimentation and things of that nature.

So there are problems which could lend themselves to models where CERN then would be an appropriate type of organization—so it's an organizing principle, if you wish.

Question: Organizing as a cosmos or an organism?

Virdee: I think people working at CERN do have respect for each other wherever they come from. The good ideas can come from anybody, and it could even be a student; sometimes it is actually a student. So this is an important aspect. Put it another way: this is freedom of expression, freedom of thought.

Another aspect which is quite important is that once you do things you also, as you go along in these complex projects, undergo peer review, which is review of your peers who are not involved in the project itself so you get external input to make sure that you're doing the right thing as much as possible. Of course, we can still make errors as such but you minimize all this by having a process of peer review and so there are certain aspects within our community which actually help in getting a product which is quite good.

Question: How should a physicist work, be, and think?

Virdee: We are trying to find the secrets of nature, so one has to be pretty modest and not sort of over-assume one's goodness, if you wish. And the other thing is when you are trying to find the secrets of nature you have to be very careful, very diligent, as I said, and always assume that there is room for error. You have to minimize that and find any possible way [so] that no error [can] be made, whether in construction or interpretation or in the analyses, and so on. So this aspect of constantly questioning oneself, whether one is doing the right thing, is, I think, quite an important one. And that allows you to understand secrets of nature when they fall in your lap.

Question: How much do you think we understand of nature? What would the picture be?

Virdee: I think one could use the phrase that if I reproduce it correctly that Newton used … he was standing on a seashore and he felt that he'd understood what was the equivalent to a pretty shell, but there was a whole ocean of truth outside, and I think that is still true. So we have made a lot of progress but there is a lot more to do. So this quest that we're on is going to continue for some time.

Notable Quotes by and about Newton

I do not know what I may appear to the world, but to myself I seem to have been only like a boy playing on the sea-shore, and diverting myself in now and then finding a smoother pebble or a prettier shell than ordinary, whilst the great ocean of truth lay all undiscovered before me.

— Sir David Brewster, *Memoirs of the Life, Writings, and Discoveries of Sir Isaac Newton* (1855)

If I have seen further it is by standing on ye shoulders of giants.

— Letter to Robert Hooke, 15 February 1676

Nature and Nature's laws lay hid in night: God said, Let Newton be! — and all was light.

— Alexander Pope, epitaph for Sir Isaac Newton

Isaac Newton

Isaac Newton (1642–1727) is considered one of the most important scientists of all time. He is the founder of classical theoretical physics and, together with Galileo Galilei, he is the founder of the exact sciences. In 1667, Newton (from 1705 on, Sir

Isaac) became a fellow at Trinity College, Cambridge, where he developed new theories about the nature of light and gravity, the movement of the planets, and other mathematical calculations (e.g., binomial theorem and differential and integral calculations).

In 1686, Newton published his book *Philosophiae Naturalis Principia Mathematica* (*Mathematical Principles of Natural Philosophy*), which laid the foundations of classical mechanics. In his major work Newton described, in mathematical terms, the law of universal gravitation, and thus combined the research and achievements of Galileo with Kepler's laws. Newton's universal law of gravitation applies everywhere, on Earth as well as in space. It is a universal scientific law.

Isaac Newton (1642–1727), painting by Sir Godfrey Kneller (1646–1723).

Newton's three laws of motion explain the reason for inertia, action, and reaction. These laws are the foundation of classical mechanics.

Newton's laws of motion were a decisive break from the traditional teachings of the followers of Aristotle, who were called the "Peripatetics". They regarded the laws of the heavens as being totally different from the laws of physics on Earth. Newton realized that there are universal laws that apply everywhere in the universe. He treated the forces as quantities that can be summed up both by experiment and mathematically. With his method, Newton laid the foundation of the modern scientific method that drives progress by coordinating mathematical and scientific theory through experimental research.

The story that explains how Isaac Newton came upon his universal law of gravitation by looking at an apple falling from a tree is another legend of physics. It goes back to "Memoirs of Sir Isaac Newton's Life" by William Stukeley. Voltaire portrays the legendary discovery in similar terms, but it is also possible that the sometimes quite eccentric Newton invented the story himself.

Newton has a central place in the pantheon of physics. His concept of absolute space and absolute time dominated philosophy and the natural sciences for more than 200 years. It was not until the arrival of Albert Einstein's theory of relativity and Heisenberg's uncertainty principle that the scientific world was again revolutionized to the same extent as by Newton's initial impact. Newton was highly honored during his lifetime. He became a Member of Parliament, Warden of the Royal Mint, and President of the Royal Society. His grave is located at Westminster

Abbey, the burial site for British monarchs. A "Newton" refers to the international unit of force (N).

Question: In the history of science, where would you put the LHC?

Virdee: I think in terms of the technology it is like the Apollo program but in terms of what might happen I would probably place it in terms of the things that were happening at the beginning of the last century when our notions of space and time were radically altered. Quantum Mechanics came into being. There could be something of that nature, so it is very important to understand what is going to happen in this energy scale, which is the energy scale of the LHC because we know that the current model—which is the Standard Model—has some shortcomings. It actually gives some nonsensical answers when you do some calculations at the energy of the LHC. At ten times lower energy it is perfect; we have not seen any discrepancy with data, so we're almost certain something will happen at the LHC. Exactly what it is, we don't know. It could lead to quite a reformation of the way we see how nature works at the fundamental level, but nature is the final decision maker; it is not us. We are trying to discover how it has solved some of these issues. We believe in some conjectures which are favored or not favored but when we do the experiments, and that is the reason we do the experiments, it is to find out. Because conjectures are insufficient we need to find out experimentally what nature has done and that's why these experiments are very important. That's why we do them. Otherwise, we could just think about things.

Question: There is a big problem: gravity. Nobody has a clue what gravity is. If the Higgs really exists, what would that mean?

Virdee: I think one has to look at what the Higgs mechanism does. What it does, is to complete the Standard Model where the photon, for example, remains massless and the carriers of the weak force, which are Ws and Zs, get a mass a hundred times of the mass of the proton. Deep down, that is the issue. We believe that it is the Higgs mechanism that breaks the symmetry, if you wish, because the simplest theories that we write down have particles which have zero mass but we know that is not true; that's not how nature has worked because we have particles of mass inside our body. We exist because it is mass that gives substance to the universe.

So something has happened. We call it "spontaneous symmetry breaking." Now, nature could have chosen another way of resolving this problem, which we may have thought of or maybe we haven't thought of so when we do the experiments we will find out the answer, and it could be the Higgs mechanism. It could be something else. In that sense—I think as an experimentalist—one has to have an open mind, although one uses guidance when designing experiments and so on.

When I and a few colleagues started CMS, we were actually trying to design an experiment that could detect anything that nature would reveal at this special energy scale. We used the Higgs mechanism because it was a very good physics benchmark which allowed many, many possibilities to be explored. If you could do a very good job on the Higgs at various masses then that detector that you designed is probably good enough for whatever nature throws at us. So in that sense, the design was made not only for the Higgs but anything that is possible at this special energy.

Now we have to decipher what nature tells us. In that sense the Higgs mechanism is just one aspect. The issue that you raise about gravity—that the Einstein theory of general gravitation and quantum mechanics are the two pillars of 20th-century science and in certain circumstances, like close to black holes, these actually give conflicting answers, contradictory answers so something is not right, and part of the problem is that we don't have a quantum theory of gravity. And there is a tension between those two pillars of 20th-century science, which are still our pillars on which the Standard Model is built. In that sense, there is progress to be made and the hope is that some of the answers, some of the results we will get from the LHC, will tell us which way to go. At the moment, for example, we are in a position where some people tell us: you have to go that way; some will tell us to go that way, some that way but maybe nature will tell us to go that way [points backwards]. We don't know.

Question: Is there anything else?

Virdee: Perhaps, there is one thing that I wish to add; the primary goal of the LHC is scientific but it has also other aspects to it which have already yielded results, like in technology. The LHC has pushed much of the technology work with industry, and so that actually helps industry to make products which are quite demanding. That's one aspect; there is always a spin-off. In our case, in CMS, I personally worked on a detector which uses lead-tungsten crystals. The light that is generated there is detected by what we call "silicon avalanche photo diodes," and these objects now find application in positron emission tomography cameras.

We sit in a very high magnetic field, and so these photo devices had to work in a very high magnetic field. In fact, people are trying out now positron emission tomography cameras inside MRI scanners to get a better, more dynamic picture of

Tejinder Singh Virdee, showing the way. (© Michael Krause)

what's happening inside the body. There are applications of that nature, and many others, in superconductivity and so on. There are also the applications of huge amounts of data distribution, which is rapid data distribution through the grid.

Another aspect, actually, is the educational aspect. Most of the people who actually come and do experiments, who work in these experiments, are from universities from all over the world. In CMS alone there are about 190 institutions. I am personally from Imperial College so I go back and teach there. All the professors who are working on the LHC go back and try to impart the enthusiasm for actually doing this kind of science into a newer generation. And in fact there are a couple of thousand postgraduate students working on CMS, on ATLAS, and other experiments, so there are a very large number of postdoctoral students. A lot of work is done by the students themselves.

And then the last aspect is cultural, because all sorts of nationalities—I've been to many, many countries, about 30 countries—you have to talk to the people there, to the rectors of the universities, the secretaries, and try to explain what you are doing and invite them to join this enterprise. So it is culturally enriching as well as the primary goal, which is scientific.

Question: Could CERN be a role model for another society, a society that is based on science?

Virdee: I think a very important goal is that these communities can lend themselves to tackling those problems. I am sure they can be solved because we have had many, many problems in constructing these experiments and the machine, and we have all overcome them. So usually one overcomes problems.

Notable Quotes by Tejinder Virdee

- We expect discoveries. This expectation is driving us—because we firmly believe that there is something that will tell us, when we discover it, how nature works and how we need to take the next step.
- The main outcome of the 20th century in physics is the Standard Model. CERN is the United Nations of physics.
- We want to find the secrets of nature, so one should be quite modest. Nature is our ultimate guide, not ourselves.

5 Dalton — Thomson — Rutherford — Bohr

The Development of the Atomic Model

For more than 2,000 years, the verdict of Aristotelean doctrine had considered Democritus' atomic model completely useless. It was not until the early 19th century that scientific research found experimental evidence that matter is actually made up of small building blocks.

John Dalton (1766–1844), an English chemist and physicist, started with weather observations and the study of atmospheric phenomena to accurately investigate the chemical properties of gases. In his book, *A New System of Chemical Philosophy*, published in 1808, Dalton laid the foundation of modern atomic theory:

- All matter is composed of atoms.
- Each element consists of identical, undivided atoms. They are responsible for the characteristics of the element.
- In chemical reactions, atoms of different elements combine to form molecules.
- In chemical reactions, atoms of one element combine in integer mass ratios (Dalton's law of multiple proportions).

Dalton's model was quickly adopted in the field of chemical research, but it was not precise and specific enough, and it could not explain the electro-physical or electrochemical reaction. Nevertheless, Dalton's method—to link theory and experiment firmly together— became the standard in physical research. In his honor, the unified atomic mass unit was named a "Dalton" (Da). One Dalton (1 Da) is approximately equal to the mass of 1 proton or 1 neutron. Today, however, it is defined as one-twelfth of the rest mass of an unbound atom of carbon–12 (^{12}C) in its nuclear and electronic ground state, and a new symbol "u" (unified atomic mass unit) has replaced the old label.

The British physicist Sir Joseph John "J.J." Thomson (1856–1940) was a professor of experimental physics at the famous Cavendish Laboratory at the University of Cambridge. His predecessors were James Clerk Maxwell (1871–1879) and John William Strutt (Lord Rayleigh, 1879–1884). In the 1890s, Thomson conducted experiments with cathode ray tubes based on similar experiments by the German physicist Johann Wilhelm Hittorf and the English pioneer, Sir William Crookes. In Thomson's experiments, the cathode ray was deflected by electromagnetic fields; he discovered that the beams had to consist of a single type of negatively charged particles "of bodies much smaller than atoms."

With Thomson's discovery it became clear that—in contrast to Dalton's atomic model—atoms are not indivisible but that they contain small, negatively charged particles. Thomson labeled these particles "corpuscles," not an unusual description for something that is not yet entirely understood. The "electron," as it was then called, was the first subatomic elementary particle discovered by man. For this discovery, J.J. Thomson won the 1906 Nobel Prize in Physics.

Together with the doyen of British physics, Lord Kelvin—William Thomson (1824–1907), credited with the Kelvin temperature scale and the formulation of the Second Law of Thermodynamics—formulated a new concept of the atom:

- The negatively charged electrons are tiny particles within the atom.
- The positive charge is distributed within the atom.

A cloud, raisin bread, or watermelons are the other possibilities that can be used in imagining Thomson's atomic model. In this model, the negatively charged electrons are distributed evenly within the atom, the pulp representing the positive charge. Thomson's model was also dubbed the "plum pudding model." This model took into account the electrical properties of matter, but it could not explain why elements generated different and unique spectral lines, i.e., why only waves with specific frequencies were sent out (the "fingerprints of the elements").

Ernest Rutherford (1871–1937) was a physicist born in New Zealand. He finished his postgraduate studies as Thomson's student at the Cavendish Laboratory in Cambridge. Since Rutherford appeared to be too young for a career in Cambridge, he went to McGill University in Montreal, Canada. At McGill University, he was able to shed light on the mysteries of the newly discovered phenomenon of radioactivity.

Rutherford managed to isolate alpha and beta rays from each other and proved that radioactivity arises from the decay of one element into another. Rutherford was awarded the 1908 Nobel Prize in Chemistry for "investigations into the disintegration of the elements, and the chemistry of radioactive substances." Rutherford is often referred to—thanks to his large theoretical and experimental skills—as "the father of nuclear physics." Einstein called him a "second Newton."

In 1907, Rutherford went to the University of Manchester, where he performed his most famous experiment ("alpha particle scattering"). In this experiment, Rutherford deflected alpha rays, which were known to be positively charged and therefore had to be made up from the "pudding" in Thomson's atomic model, from a very thin (0.000004 cm) gold foil. This gold foil was about one thousand atoms thick. Behind that foil there was a circular detector film of zinc sulfide. If the mass of the atom was evenly distributed as in Thomson's model, all alpha ray particles—consisting of double-charged helium ions—should pass through the film unhindered. In Rutherford's experiment, however, it turned out that some of the particles were distracted. Some of them were even thrown back.

With regard to his gold foil experiment, Rutherford said, "It was quite the most incredible event that ever happened to me in my life. It was almost as incredible as if you had fired a 15-inch shell at a piece of tissue paper and it came back and hit you."

In 1911, Rutherford published an article about the results of his scattering experiments, titled "Structure of the Atom." In that article he described his model of the atom:

- The atom consists of a positively charged, extremely small and extremely solid core.
- Negatively charged electrons move in a circle around this core.
- Between the nucleus and electrons there is empty space.

Rutherford calculated that the diameter of the nucleus was less than 3.4×10^{-14} meters. It was known that the gold atom had an overall diameter of about 3×10^{-10} meters. The nucleus therefore was about 10,000 times smaller than the entire atom!

Rutherford developed an atomic model in which the electrons—much like the planets orbiting around the Sun—orbit around the core of the atom. His planetary model superseded Thomson's pudding model. Rutherford's model was very clear, but it had one crucial flaw. In his model, the electrons could take any possible orbit around the nucleus. A consequence of this would have been the fact that one single element could have completely different properties. With this model, it was also impossible to explain the fact that different elements had different spectral lines. In addition, the orbiting electrons would—according to the laws of electrodynamics—constantly be losing energy and therefore they would crash into the nucleus. Although Rutherford's atomic model was a huge step forward, it produced unstable elements—which obviously was not right.

Niels Bohr (1885–1962) was one of the most influential physicists of the 20th century. Bohr, aided by Max Planck, Albert Einstein, Werner Heisenberg, and Erwin Schrödinger, put together the foundations of quantum theory, which is the basis of modern atomic physics. Bohr, coming from Danish nobility, had studied in Copenhagen. He then furthered his postdoctoral studies with J.J. Thomson in Cambridge and with Ernest Rutherford in Manchester, respectively. In 1913, Bohr further developed Rutherford's atomic model by adding concepts from Max Planck's quantum theory. Bohr's model did not arise from experiments, but was proven by actual existing properties of the hydrogen atom.

Bohr's Atomic Model

- Electrons orbit the nucleus exclusively on circular or elliptical orbits.
- The electron orbits are located exclusively on certain quantized energy levels.
- Transitions between the electron orbits are only possible in jumps (quantum jumps) caused by a simultaneous absorption or release of energy.

Although Bohr's atomic model was able to accurately describe the spectral properties of simple atoms such as hydrogen, in more complex atoms, it was not applicable. In 1921, Bohr eventually developed the "structural principle." It specified the structure of electron trajectories: "the electrons in the atom are arranged in distinctly separate groups, each containing a number of electrons equal to one of the periods in the sequence of the elements, arranged according to increasing atomic number" (Atomic Structure, *Nature*, March 24, 1921).

In Bohr's model, the electrons move in certain orbits of fixed, or "quantized," size and energy, dubbed "shells." Only the outer shells determine the chemical properties of the atom. Bohr's model thus provided a theoretical explanation of the chemical elements. It is still valid today—with some modifications applied through the cooperation of Werner Heisenberg—as the basis of modern atomic physics. The 1922 Nobel Prize in Physics was awarded to Niels Bohr "for his services in the investigation of the structure of atoms and of the radiation emanating from them." In 1921, he founded the Institute for Theoretical Physics in Copenhagen. In the following decades, Bohr's institute became the center and starting point of international nuclear physics research.

6 The Man Who Built the LHC: Lyn Evans

Lyndon Rees Evans.
(© Michael Krause).

Lyndon Rees Evans was born on July 24, 1945 in Aberdare, Wales (UK). Evans began to study chemistry; then he switched to study physics at the University College of Swansea. His doctoral thesis dealt with plasma formation in gases at high laser irradiation. Since 1969, Dr. Evans has worked at CERN. In 1971, he joined the team for the 300 GeV project, later renamed Super Proton Synchrotron under the direction of John Adams. Evans was responsible for the conversion of the SPS into a proton-antiproton accelerator, the first ring accelerator of this type worldwide. He is an expert in the design and construction of particle accelerators and worked in the US towards the completion of the first superconducting storage ring at Fermilab near Chicago. From 1988 to 1993, he worked at the Superconducting Super Collider in Texas. At CERN, Evans became leader of the SPS in 1989, and later head of development for the Large Electron Positron (LEP) accelerator. In 1993, "Evans the Atom" was appointed project manager for the construction of the Large Hadron Collider. In 1995, Evans and his team presented the LHC conceptual design, a detailed baseline for its construction. Since 2001, Dr. Evans has been a Commander of the British Empire. He is a Fellow of the American Physical Society and a Fellow of the Royal Society.

The LHC tunnel. (© 2012 CERN AC-120617008)

Question: What is the LHC?

Evans: The LHC is the Large Hadron Collider. A hadron is a proton or a neutron, the nucleus of a hydrogen atom. It is a particle accelerator which, with very high energy, collides two proton beams and produces new particles, converts energy into mass, basically.

Question: What is your job here?

Evans: I am the project leader. I've been working on the project since 1993.

Question: How do you feel about the whole project?

Evans: It is very exciting now. Normally a project lasts four or five, maybe six years. This project is so immense and it has taken so long to finish. It is a very exciting time now for us and a very hard time, too.

Question: How did you get involved in the LHC?

Evans: That's a long story. I have been at CERN for a very long time. In fact, I arrived in 1969, and I worked on building accelerators all my career. When the

LHC was ready for approval by the council, I was appointed as the project leader. In 1993, first of all, and it was approved in 1994, and I've been running it ever since.

Question: Are you proud of your achievements?

Evans: I think the LHC is the future of CERN and of particle physics. This machine is going to be the flagship machine for the next 20 years. I think there is nothing bigger than that a scientist could hope to achieve.

The Large Hadron Collider (LHC)

The LHC is a synchrotron, like its pre-accelerators, the PS and SPS. In synchrotrons, a synchronized, high-frequency alternating electric field in the microwave range is used to accelerate particles. In contrast to the pre-accelerators, protons in the LHC are accelerated in two parallel, but separate, beam pipes. The beam in one pipe circulates clockwise while the beam in the other pipe circulates counter-clockwise. When the protons reach their maximum energy, the two beams are brought into collision. The LHC has a circumference of 27 kilometers. It is the largest and most powerful particle accelerator in the world.

The protons come from a simple bottle containing liquefied hydrogen gas. The hydrogen atoms are stripped from their electrons by an electric field. Then the protons are accelerated by a linear accelerator, LINAC2; then the "kicker magnets" inject the particle beam into the circular accelerators. These magnets produce a deflection field, causing the beam to deflect into the trajectory of the LHC. Once in the LHC, 1,232 dipole magnets keep the particles on their circular path through their strong magnetic field of up to 8.3 Tesla.

These magnets are cooled down to 2 degrees Kelvin, or 271.3°C, by superfluid helium (He II)—a temperature colder than outer space. The magnets thus become superconducting, while the electrical resistance drops to zero and a much higher current is able to flow without resistance. The acceleration of the protons occurs on acceleration tracks with the help of "cavity resonators" in which high-frequency electromagnetic waves (in the microwaves range) are applied. During this process of continuous acceleration, the protons "surf" on the electromagnetic waves and thus get faster and faster before they collide with maximum energy in the detectors (e.g., the ATLAS, CMS, etc.) and decompose quickly into other particles.

Particles in the beam move almost at the speed of light (99.9999991%). The total volume of the LHC beam complex that has to be cooled down is about

9,000 cubic meters. Within the beam pipes, a vacuum of 10^{-13} atmospheres is needed. The vacuum prevents the moving particles from colliding with gas atoms. Despite all of these measures, the particles during their "journey" can accidently collide with the remaining atoms so the beam must be focused continuously. This is done with strong quadrupole magnets (magnets with four poles). The LHC features approximately 500 of these magnets, each about three meters long. The quadrupole magnets as well as the dipole magnets are superconducting.

In the experiments, up to 600 million proton-proton collisions per second occur. Besides the usual hydrogen nuclei, lead-lead collisions are also analyzed (in the ALICE experiment). The two particle beams consist of 2,808 bunches each. The minimum distance between bunches is about 7.5 meters (25 nanoseconds); bunches contain about 10^{11} protons each. The nominal energy of each proton is at collision point 7 + 7 TeV (2010/11: 3.5 + 3.5 TeV; 2012: 4 + 4 TeV). The number of turns of each bunch is about 11,245 per second. The collision rate (luminosity) is about 10^{34} collisions per square centimeter per second.

Question: Is there something that really motivated you to become a physicist?

Evans: Well, no single thing. If you are interested in science, you are interested in science in general. In fact, I went into university to study chemistry, not physics, but after my first year I found that physics was much more interesting for me.

Question: You said that this is the flagship of physics. Why?

Evans: There isn't anything else like it. I think that it moves the frontier of high energy by a very large amount and there is no doubt that there will be some very fundamental discoveries made with this machine so it's very exciting moving into a completely new regime where nobody has ever been before.

Question: What do you think they are going to find?

Evans: There is a whole shopping list of what is expected. We all know that the Higgs boson is one element. If the Higgs boson exists then it has to be in the range of this machine to find it. There are other issues which are on the shopping list of questions to be answered. But always with these kinds of new studies, in a completely new regime, you can expect surprises and I think the most interesting things are the surprises.

Question: What are the ground data of the LHC? How big is it? How much power do you have to put in to make these beams exist?

Evans: Well, paradoxically, less power than we have ever used before at CERN. In our conventional accelerators ... to give you a number, we use about 200 megawatts of power. But the LHC is a superconducting machine. It's made with superconducting magnets that consume no power at all. In fact, the power consumption in the LHC is in the liquid helium refrigerators that they use to cool the magnets, not in the magnets themselves. The LHC consumes about half of what we used to consume in the old machines for a much higher energy.

Question: How much time did it take to build the LHC, and how many people were actually involved?

Evans: The LHC was proposed in 1984—in fact, in a meeting in Lausanne. The project was approved in 1994 so that was ten years from the first conception to the approval and it's taken more than 13 years to build. So it's a very long project in itself.

Question: How much did it cost to build the LHC?

Evans: The final cost will be about three and a half billion Swiss francs—a bit more than two billion euros.*

Question: How many people were involved?

Evans: That's a very difficult question to answer. I think the CERN staff working on the LHC has been about 400 but [the] contractor's staff and staff in other laboratories is not included in this number, because part of the LHC was also built outside of CERN: in the United States, in Russia, in India, in Canada, in Japan. I don't even know how many people were working there so it's going to be a good thousand people working on the LHC at any time. And that's only on the machine. On the detectors—that's another story.

Question: How is the beam generated?

Evans: Of course, the LHC uses the existing infrastructure of CERN: the old

---

* Data for overall project expenses for the LHC vary. According to the CERN factsheet, the actual costs are more likely to be more than six billion Swiss francs. Research and development, and tests are included in this calculation but not the costs of collaborations.

accelerators. In fact, the beam starts from a hydrogen bottle. Hydrogen gas is ionized, producing protons, which are positively charged nuclei of hydrogen, which is then accelerated first of all in a small linear accelerator. Then it goes through three other accelerators, increasing the energy until it is injected into the LHC. This entire infrastructure already existed at CERN. We didn't have to build that. If we had to build the LHC on a green field site where we had to build everything, the cost would have been much more than the 2 billion euros that I mentioned.

Question: The beam itself, how does it look? How big is it?

Evans: The beam itself consists of bunches of particles traveling at the speed of light, very close to the speed of light. Inside these bunches, which are a few centimeters long, there are trillions of particles, of protons. The size of the beam is about one millimeter inside this vacuum chamber except at the experiment's collision points where

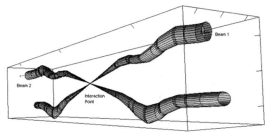

Beam sizes at the ATLAS detector. (© 2012 CERN)

we focus the beam down to about 15 microns. Fifteen microns is about half the width of a human hair when we bring the beams into collision in the detectors.

Question: What does the beam look like?

Evans: We have monitors, but you can't see it with your own eyes. We have TV screens that we can put in and we can see an image of the beam spot—a very bright image of the spot. We have other detectors where we can measure the beam profile; we can see how wide it is. And everywhere in the machine we can measure the beam position because we have to keep the beam in the center of the vacuum chamber all the time so we have lots of diagnostics that allow us to see the beam indirectly.

Question: How dangerous is the beam?

Evans: The beam contains a lot of energy. The full power of the beam is 350 megajoules, which doesn't mean much to you but it is as much as 80 kilos of TNT—maybe that means a bit more. So it's got to be treated with respect. I think if we lose the beam in the accelerator, then it's going to drill a hole through the vacuum chamber. One important thing is the systems that keep the beam safe in the machine, very sophisticated. When we don't need the beam anymore we extract it and we send it to a specially designed beam dump, a block, an absorber that can take the power of the beam safely so it is a very powerful beam, sure.

Question: How could one imagine that?

Evans: As said, 80 kilos of TNT. I have never seen an explosion with 80 kilos of TNT but I can imagine it.

Question: What is the equivalent, or amount of energy, when the beams collide?

Evans: When the beams collide—of course most of the time, even though the beams are very dense, nothing happens. They just pass through each other without interacting. But a small portion of the particles collide with each other and that gives collisions of 14,000 GeV. That's about ten times higher than the energy ever achieved before by man. Cosmic rays coming from outer space have higher energy than that but these events are much rarer than in the LHC.

Question: Can we see cosmic rays sometimes? How can regular people find out more about them?

Evans: Cosmic rays are going through our body all the time and if you go into our exhibition, the "Microcosm," then you see detectors which are recording these cosmic rays as tracks. We don't feel them. They go through us; they are bombarding us all the time.

Question: Where do cosmic rays come from?

Evans: They come from the cosmos. It is not completely understood how they are made but they are bombarding us from outer space.

Question: Is the product in the LHC similar to cosmic rays?

Evans: The cosmic rays, you can only detect them by chance. They are not predictable. But in the LHC, the collisions will be in a very precise position in the center of the ATLAS detector and ATLAS will be able to record all of the debris coming out of the collisions to reconstruct what happened during the collision itself.

Question: What will be the next step, after the LHC?

Evans: The LHC will tell us what to do next, I think. We will make discoveries at the LHC that will guide us—as always in science. You do an experiment which should guide you to the next step.

Now the LHC is a compromise because when you collide protons together ... a proton is a composite object. It is not a fundamental object like an electron. An electron is an absolutely fundamental object. A proton, I make the analogy sometimes, is like an orange. It's got pips inside; it's got three quarks inside the proton. When you collide two oranges together then you will occasionally get a collision between the hard pips but you will always get the pulp. So, in the LHC, there is this very large background, there is the pulp, and very occasionally the quark-quark collisions that give the really fundamental interaction so the LHC is a dirty machine.

You are looking for the quark-quark collisions among the noise of the background, the pulp of the proton. So the next machine should be an electron collider but we don't know how to build electron colliders like the LHC, going around in a circled orbit, because electrons emit light when they are bent, and they lose energy. And the higher energy you give to them, the more they lose. So the next machine has got to be a linear collider. It's got to be two electron accelerators colliding each other in a straight line and I think the LHC will tell us what energy this machine should have. This study is called the international linear collider, which is a study now preparing the next step after the LHC.

Question: Do people already work on about the next steps?

Evans: Very much so, yes. I think that the DESY laboratory in Hamburg has been on the forefront of the R&D necessary to make the technology of this to work. I think it's waiting now in the starting blocks, waiting to see what the discoveries of the LHC are, what energy the machine should be, and it should be given the go ahead to start construction around 2015.

Question: The LHC and ATLAS, what is the connection?

Evans: ATLAS is one of the two very large detectors, general-purpose detectors, which is designed to cover all aspects of the physics of the LHC. The other one is called the "Compact Muon Spectrometer," CMS.[†] And there are two smaller detectors as well, which are specifically designed for colliding heavy ion beams, because we can also accelerate heavy ions to produce the quark-gluon plasma, the very beginning of the Big Bang. And there is another, smaller detector which is

---

[†] CMS—Recipe for a Universe: Take a massive explosion to create plenty of stardust and a raging heat. Simmer for an eternity in a background of cosmic microwaves. Let the ingredients congeal and leave to cool and serve cold with cultures of tiny organisms 13.7 billion years later." (cms.web.cern.ch)

designed to look at some of the fundamental problems of why matter exists and there is no antimatter in our universe anymore. But ATLAS and CMS are the really big ones, and ATLAS is the biggest of them all.

Question: How close is the collaboration between ATLAS and the LHC?

Evans: You cannot have ATLAS without the LHC and you cannot build the LHC without the detectors so it had to be a very close collaboration. Of course, the way the two are built is different. The LHC machine is built by CERN, with some cooperation at a 20% level from other institutes that I mentioned, but the detectors, and in particular ATLAS, is built by very large international collaborations, where CERN is only a 20% contributor. So it's a different construction completely but it's very important that we have a very close collaboration, and this has been the case all the way through construction.

Question: What has caused the delay in 2008?

Evans: I think with this machine...nothing like this has ever been built. It's no surprise that there are problems along the way. One of our problems was a fault in one element which was supplied by one of our collaborators—so that's a little embarrassing—which we had to fix. But I must say that there has been only small problems, and actually in a 13-year project, a few months delay is not a big deal, and the machine is going to run for another 20 years.

Lyn Evans was announced as the new Linear Collider Director on June 20, 2012. Dr. Evans will lead planning and design for the new linear collider and its accompanying detectors. The ILC will accelerate particles to between 0.5 and 1 TeV and will use superconducting technologies.

Notable Quotes by Lyn Evans

- The LHC converts energy into mass.
- The LHC is the future of particle physics. This machine is going to be the flagship machine for the next 20 years.
- When the beams collide, of course, most of the time nothing happens. They just pass through each other without interacting.

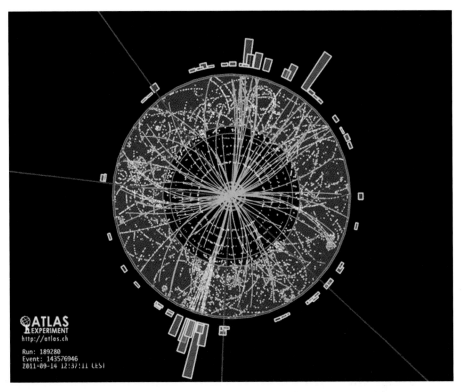

An event with four identified muons from a proton-proton collision in ATLAS. The muon signals are the red lines from the center to the frame of picture. They are a possible signature for Higgs particle production. (© CERN-EX-1112301-01)

7 Physics, Music, and Art: Tara Shears

The LHCb Experiment

Tara Shears and Sebastian White meet for a coffee break on the terrace of the CERN cafeteria. From here, Mont Blanc can be seen in good weather conditions, i.e., sunshine and clear blue skies, so only on a few days per year. Thus Mont Blanc is similar to dark matter—it is there certainly, but very difficult to take in.

White: There is a rumor that Mont Blanc is somewhere in the distance there.

Shears: Although we can't see it—somewhere, not today. It's too hazy. But it's a gorgeous day.

Sebastian White and Tara Shears on the terrace of the CERN cafeteria. (© Michael Krause)

Question: We are here at the cafeteria. What is so special about it?

White: There are actually more elegant cafeterias around the world even at scientific places.

Shears: Oh, there is no cafeteria like this one, because this is where we really do physics. I think this is where everything that is going on at CERN has its roots; somewhere in the cafeteria, and over a coffee, I think.

Question: What kind of talk do you mean?

Shears: Important is the moment you have overheard the talking at the other table.

Question: What do you mean?

Shears: Very often when you are thinking about a problem you get stuck in it and you can keep going around in the same circle for hours, for days, for weeks even but all you need to do is to get out of it somehow and the best way to do that is to sit around the table with your friends. Talk about something completely different. Talk about mansions, talk about skiing, then your mind goes just back to the problem and something is happening; you have a real connection, subconsciously. One of you says something; another picks up and says something, and it's like around the table suddenly this problems drops into place, like a jigsaw puzzle.

Question: So the CERN cafeteria is more like the great subconsciousness collider?

Shears: If you like. It's just a very good place.

White: You could also say, "stream-of-consciousness collider," if you like.

Shears: It is a very good place to relax and work things out. That's why it is an important place. And you can talk to people really freely here. You can talk to theorists, you can talk to experimentalists, making experiments apart from your own. You know, we do talk to each other. It's a good place to do it. Very civilized. Nice coffee.

Tara Shears is a theoretical physicist. She was awarded a PhD by the University of Cambridge. Today, Dr. Shears is a reader at the University of Liverpool. In 2000, she started research on the production of quarks at the Tevatron (Chicago, USA) as a part of the CDF experiment (Collider Detector at Fermilab); at the time, it was the particle accelerator with the highest energy level in the world. Dr. Shears then moved to CERN for further research on B-quarks and the production of W bosons in connection with the OPAL experiment (LEP), and then she later moved on to the ATLAS experiment. Today, Dr. Shears works within the LHCb experiment, which mainly studies the causes of the uneven distribution of matter and antimatter in the universe (CP violation). Her research focuses on tests of the Standard Model in the field of electroweak force. Tara Shears is a member of the CERN working groups QCD (quantum chromodynamics) and Exotica physics. In her leisure time she is very much interested in poetry and classical music (http://twitter.com/tarashears).

Question: What kind of place is this? What kind of connection do you have to this place?

Shears: This is the CERN Council Chamber. It is the room where the member states of CERN meet to discuss, for instance, the finances they are going to pay to keep CERN going. It's where the CERN Council meets to discuss strategy. It's where our experiments meet to discuss our latest results among ourselves. It's where we have theory seminars, it's where we hear updates on the accelerators—so discussion of every level of CERN operation, from the everyday things of physics right up to the top strategy among the high-ups, goes on in this place. I am an experimental particle physicist here, one of thousands, and I come into this room

Tara Shears at the CERN Council Chamber. (© Michael Krause)

really most often to listen to what's going on in physics, the results of the other experiments, or within my own experiment. We have weeks of meetings every year, some of which occur in here where we update each other on what's going on and what we found.

Question: What is your job here?

Shears: I work on the LHCb experiment, and I analyze data. Although I officially work for the University of Liverpool, I am associated to the experiment out here. I come out to analyze the data. I show my results to all the experts here, I discuss it with them. You really have to think about our experiments as a distributed system where a whole load of people are working together but not in the same room. Normally, we are spread around the world but every so often we have to come together to talk face to face about what's going on; because it's by far the easiest way to do things. You just can't do it every day. I come out to CERN when I have an extended period of work to get through, to analyze data, but also when it's important to sort of touch base if you like—to talk to people and really find out what's going on and show them what you've been doing.

Question: What is the communication like? What kind of data do you exchange? What are those meetings for?

Shears: It seems so natural to us and perhaps it's quite strange if you're outside the system, because the way we work is quite inefficient in a sense. In my experiment, there are 700 of us all together and we all play our part in collecting the data, analyzing the data, and as such we will have a good say in the data. If you look at any of our papers, we all appear in alphabetical order on the author's list. There is no hierarchy on the outside when it comes to priority in the results. Now to get there it means that we have to work in a very rigid structure. We can work with colleagues on analyzing a particular piece of data, trying to understand something, for example, trying to understand the behavior of antimatter.

We will analyze data with computers, and we'll compare data to our theory. We look at how well those comparisons are behaving, and then when we're happy with it inside our physics working group, then we show it to the rest of the experiment and see what they think. And that's an extended process of peer review that can go on for a month—something like that, and it can be quite severe until everybody is happy or is satisfied with the way you analyzed the data; that it's correct. When that's OK, then we release our results publicly. We show them at conferences and that's the stage where we'll present them to the other experiments here or sort of elsewhere.

And then we will turn it into a publication. That involves peer review from thousands of other particle physicists from around the world so it's quite a slow process, and it does involve in every step keeping people informed about what we're doing because that is the only way to work with so many people. We have meetings through the Internet or weekly video conferences, and that is how we stay in touch. We also make visits to talk to each other; in that way you keep a personal aspect of what's going on. And then after that we have discussions via phone, via Skype—you name it. There is a lot of discussion face to face, and vocal as well. It's the only way really to keep everybody involved, and everybody sure of doing the right thing.

Question: What is the right thing?

Shears: The right thing is what we are trying to do, and it's very difficult to arrive at a conclusion. We are looking for answers to difficult questions—we are looking to understand the nature of antimatter, and we are looking to discover more about the universe at the very smallest level, the behavior of the fundamental particles that make it up. And we don't know what the right answer is. But what we do know is … we have a theory but that only describes the best estimate of the right answers. And we have data that might give us insight into what the right answer might be.

Our job is to analyze the data, try not to bias ourselves; do it objectively as far as we can, compare it to what we know, and try and see further. Try to see the discrepancy. And it's in that discrepancy that we have a window, if you like, onto what exists in the universe beyond our current understanding. That is the approach we try to take all the time but it is difficult because we don't know what the right answer is so that's why it takes cross-checks by ourselves, by our colleagues, by other experiments, before we really have confidence that we are seeing anything beyond what we know already.

Francis Bacon (1561–1626)

"Aphorisms concerning the Interpretation of Nature and the Kingdom of Man," or: "Novum Organum Sciantiarium," were published in 1620. Bacon's main opus was instrumental in the historical development of the scientific method.

• Man, as the minister and interpreter of nature, does and understands as much as his observations on the order of nature, either with regard to things or the mind, permit him, and neither knows nor is capable of more.

- The unassisted hand, and the understanding left to itself, possess but little power. Effects are produced by the means of instruments and helps, which the understanding requires no less than the hand. And as instruments either promote or regulate the motion of the hand, so those that are applied to the mind prompt or protect the understanding.

- Knowledge and human power are synonymous, since the ignorance of the cause frustrates the effect. For nature is only subdued by submission, and that which in contemplative philosophy corresponds with the cause, in practical science becomes the rule.

- Man, whilst operating, can only apply or withdraw natural bodies; nature, internally, performs the rest.

- It would be madness, and inconsistency, to suppose that things which have never yet been performed, can be performed without employing some hitherto untried means.

- The sole cause and root of almost every defect in the sciences is this; that whilst we falsely admire and extol the powers of the human mind, we do not search for its real helps.

- The subtilty [sic] of nature is far beyond that of sense or of the understanding: so that the specious meditations, speculations, and theories of mankind, are but a kind of insanity, only there is no one to stand by and observe it.

- It is in vain to expect any great progress in the sciences by the superinducing or engrafting new matters upon old. An instauration must be made from the very foundations, if we do not wish to revolve forever in a circle, making only some slight and contemptible progress.

Question: Is there something like common sense in your community?

Shears: There is a type of physics intuition that you develop as you carry on doing analyses if you like but that's based on a prejudice and the prejudice that we've always adopted is that the universe should really be simple, and that it should be understandable. Now, prejudices serve us really well but understanding as much as we have has really guided our way. But as to whether that is the way that the universe is, we don't really know at all. It's one approach, let me say that.

Question: Is the universe beautiful and simple? Or is it inhumane? What is it?

Shears: We would like to think it is simple and beautiful but at times it feels absolutely inhumane. We are trying to understand it. By "beautiful" we do simply mean our fascination and by "simplicity" we mean an elegance in the sparseness of the description you need to describe so much of it. And the great power of describing the universe in terms of our theory is that it is in part a very simple theory. The idea is very simple, and this very simple idea has such far-reaching consequences and is able to describe so much of the universe. To us, that is simplicity; that is wonderful; that is a beautiful thing—the illumination that this very simple idea gives you by linking together seemingly disparate stuff. Electricity, the weak force that governs radioactivity, for instance, is bringing together things that seem so different in everyday life. Of course it's a simplification but there are not as many different things in the universe as you might have thought and that simplification gives you a feeling that you are touching something very deep and fundamental in the universe. And it's that feeling that you are somehow able to glimpse the skeleton under the universe that you see around you which is so compelling if you are a particle physicist.

Question: What kind of feeling is that?

Shears: It is a very humble feeling because there is no reason on Earth that we should be able to understand the universe and there is no reason on Earth that we should be able to understand and devise a theory that describes so much of what we see experimentally. I find it breathtaking in a very humbling fashion. I find it a very good feature of mankind that we are able to work together and literally work together throughout the world to make this happen, set aside political differences, set aside for the moment cost issues—thinking that this is a very important thing to do—to try to understand the universe and work together, putting our egos aside to come to as far as you can in the understanding of how it works.

I think all of those aspects about it: the simplicity of the universe that makes you feel insignificant against it. But that it's a wonderful, wonderful thing and seeing these deep connections—not even seeing, but feeling that they are there is a feeling so wonderful when you strive so hard to understand it, that it makes you so compelling that you want to understand it again and understand it better. All of these things together, I think, make particle physicists who want to work and stay in the subject absolutely compulsive. We all love what we do and that's why we do it.

Question: You say that it is a kind of humble feeling. Is it like admiring something, because the universe is so big and simple in a way?

Shears: There is always a sense of wonder at the scales but, no; it's more of a feeling in front of great art or when you listen to great music that moves you, and moves you away, takes away all your concerns from everyday life and just usually with a single feeling that something is ephemeral, or that you are sort of empty in comparison to something absolutely wonderful. It's more that feeling when you look at the equations that we have derived to try to describe particle physics—when you think about what they actually mean, when you think about the universe in terms of this very simple structure, then it is an admiration that you have. It is a humility that you have in front of it. The feeling you have when you consider it is very much like being enveloped in a great musical experience, I find. It removes you from yourself. It removes you from your everyday concerns; it removes you from any feeling that you have that you should be able to understand it. It's beyond you, completely beyond you. In that sense it's a bubble beyond you; you can only try and do your best but the very fact that you've had this little human offer to you gives you a lifted soul.

Question: Do you remember the first time that you had this feeling?

Shears: Oh, my gosh, I think if you are a particle physicist you always have that feeling when you understand anything. It's part of the whole process of seeing deeper and seeing further into anything.

I remember when I was at school and the very first shock I had about the world around me was—I remember this distinctly, I was five years old, and in my primary school class we were being told about the Solar System. Now up to then my world consisted of my house, my family, my school, and the immediate environs around it. I had no concept for anything apart from this existed—let alone the other side of the planet, and now to hear about the Moon and the Sun! Of course I knew that there was a sun, but also the other planets; somehow at that moment, it gave me such a shock to realize that there were other immense planetary bodies around and that it wasn't just the Earth. It must have been in microseconds what the world of physicists went through when it went from an Earth-centered universe to a sort of the Sun being the center of the Solar System. All of a sudden you realize that in some sense the universe doesn't revolve around you, that there are bigger things out there and that is the first time I think I felt it. And it was such a shock that I still remember it. But there is something of that feeling in solving every single problem that you come across and that's still the feeling that I have in particle physics today.

Question: When did you decide to become a particle physicist? When was the initial moment when you said, "Wow, this is it"?

Shears: That's something that grows on you because when you are young you don't know so much about particle physics. You are interested in so many things and you want to do them all, and they all seem a great way to spend the rest of your life, but what happened to me was that I studied physics at school among other things. I was split when I went to university whether I wanted to study English or study physics so I picked physics, thinking that I could always go back and enjoy literature whereas physics is something that you have to learn because of the mathematics you need to describe it.

What I really enjoyed about physics was understanding systems as completely as you can and that means understanding in terms of what makes them up and what those little bits that make them up are doing and what makes them behave. And that's sorts of a natural progression down to particle physics because you can't get smaller things than fundamental particles, and the way these fundamental particles are linked together gives you the structure of the universe. It's the simplest way if you like to try to understand what's around you. You just go down to what it's made of. And I found that so interesting. So then I did a PhD in particle physics, and then I decided that I wanted to keep on going and wanted to try and understand more, and I am here, still trying to understand more.

Question: Another thing is that we know that we only know about 4–5% of what is out there. What is your feeling about that?

Shears: Well, it's that feeling of being very small again. Gosh, there is so much more to the universe than I ever, ever thought. What on earth is this stuff? What is this dark matter that makes up 23% of the universe? What on earth is this dark energy? What's making the universe behave like this? And it becomes a source of wonder rather than anything else and rather than putting you off and making you think, "Oh, this is rubbish!" the universe is always going to throw more stuff at you that we are not going to understand. It makes you want to know more. It makes you want to understand more about the unseen side of the universe and just find out how it ticks. And this compulsion to understand never stops.

Question: What are the ingredients for a good particle physicist?

Shears: I used to think that to be a good particle physicist you have to be an extremely good theorist and mathematician to cope with the equations, and it's important because you need to have a certain mathematical ability. But what I have grown to realize is that it is perhaps more important to be stubborn, insistent, because there are really dark moments when you are a particle physicist; when you have data in front of you and you don't understand them. And that this is a part of the universe nobody has written a book about—it's new. It's your job to find out what's happening in it. It takes time for you to understand it and to be sure that you understand it. You have to devise the questions to ask to make sense and you have to devise how you are going to test that; you are making sense of the data, and cross-check, and cross-check, and cross-check until everything is consistent. Then when you converge on a consistent understanding, then you announce it to your colleagues.

Now, very often, particularly when you start off in particle physics, you don't know which way to start off analyzing the data. You really have no idea. You have to take a path, follow it down—most often it goes nowhere. You have to take another one, a different approach to the data. Try and look at it in a different way and that will be going to nowhere, too, but persistence is the key. You have to keep on going, keep on going, and not be put off. You have to be at a level where you find yourself waking up at two or three in the morning with new ideas on how to analyze the data and then be prepared to write them down before you forget them, which normally happens. And that is the most important ingredient I think. This desire to always understand what takes you beyond the bad times when it's so hard to understand. And without that, without that desire to do it, you are not going to be up to pushing your understanding forward. You'll never find answers.

Question: Where are we now? What kind of frontier are we facing in particle physics? And what is beyond that frontier?

Shears: I think it's very interesting. We are in a situation where … I think it's a bit like physics at the end of the 19th century where you had eminent physicists announcing that there is no more to be found—that we had understood everything. And all of a sudden something came along that was quite different to contemporary theory and turned everything on its head. In a sense, we're at a similar sort of position although we know it's not the full story. We have this wonderful theory of particle physics that we have not managed to disapprove anywhere but we haven't managed to prove that it's entirely correct either but the one thing that we do know about it—besides reflecting what we see in our experiments incredibly well—is that it's not the whole story of the universe by any means. It doesn't describe everything that we know about the universe; about gravity, for example, that is not contained. And it doesn't explain other things; it doesn't explain phenomena, which if you are going to have a theory, a very fundamental theory describing the universe, that should really give you insight, which this one doesn't.

It is like we are having a glass wall in front of us. We are in a building; you are in a room with a glass wall that you walk up to see if you…you want to go beyond it. You haven't found where the door is because it's a hidden door. Somewhere there has to be a crack. Somewhere you need to find a way of finding what's beyond there and we are at a stage of somewhat hammering on this glass window—anywhere we can think of to just try to find where the weak point is. And when we've done that then we only can step out beyond where we are now and see what lies beyond. That's what I hope we'll do.

"There is nothing new to be discovered in physics now. All that remains is more and more precise measurement."
— William Thomson (Lord Kelvin, 1824–1907)

"An eminent physicist remarked that the future truths of physical science are to be looked for in the sixth place of decimals."
— Albert Abraham Michelson
(1852–1931, "Michelson-Morley" experiment)

Question: What kind of things could be behind that wall of glass?

Shears: We think of those things all of the time. We have so many theories [about] what might exist over there, but the trouble is they are only theories, only ideas. They need experimental proof to become fact, or at least to become scientific law. There is no shortage of visions if you like what the universe looks like out there beyond this glass window but the problem is we don't know which one is correct—for that we need to run experiments. We need to test them and those who survive that process and that yield predictions that we can test, and if our tests agree with … well, then that is the vision that will be unveiled when we've got through.

"When it comes to atoms, language can be used only as in poetry. The poet, too, is not nearly so concerned with describing facts as with creating images."

— Niels Bohr, *Physics and Beyond*, 1971

Question: What are the main questions in physics today?

Shears: The main interesting questions in physics today all have to do with the universe and the way it behaves and the way it operates and try to find out more about it. For me, the most interesting questions are really what lies beyond our current understanding. What is the universe really like? Are there such things as the fundamental particles that we are studying, or should we be looking beyond that to something else? What is antimatter?

My experiment specializes in trying to understand what antimatter is, and why is it a little bit different to normal matter? Bizarre stuff that we don't see in the universe now but which have made up half the universe at the Big Bang, and what behavior is responsible for the way the universe evolved to allow us to be here to ask the question? That's an important one for me. Almost anything you can think of about the universe's behavior, we keep on asking why it happens, why it happens, why it happens. You get down to one of these big questions in physics and it almost invariably involves questions particle physics is trying to address. So there is the question of why antimatter is not here, and there is the question of what the universe really is made of at the deepest fundamental levels, and why it should be so.

Question: Where has all the antimatter gone?

Shears: We don't know. We built experiments to try investigate it, and at CERN there's the LHC that can make matter and antimatter that you can use to investigate but there are other experiments like the ALPHA experiment that makes samples of anti-atoms directly and stores them for about 15 minutes so far so you can look at their properties directly and then compare them to normal atoms. So you try in so many approaches as you can. It is part of this whole problem solving I talked about earlier. You try find as many directions through the problems as you can and see whether you can come up with a consistent answer at the end, and if you can then it's a strong argument for being on the right track.

Tara Shears—thoughtful. (© Michael Krause)

Question: Which of these three things symbolizes the universe best: onion, pomegranate, or walnut?

Shears: I would think that inside the walnut there is a very complicated structure that could almost be fractal with faults. I would think that in the onion there are certain layers that we can't get beyond. And this, this is a different model, indeed, almost having bundles of the universe inside—many, maybe multiverses.

As a true scientist, I am not going to pick any of them, because I literally do not know. That's the trouble with experimentalists. You are very reluctant to place any bets on anything.

To me the universe is a bit like one of these Pythagorean ideal shapes so we have this conception of the universe that we try to make simple, simple, simple. And the universe at its heart, we believe, is one of these perfect shapes, perfect sphere, something like that. This is what we want the universe to behave like; this is really what we want it to be like. We find that when we start making our measurements that it's not quite like that and we have to make this picture of this perfect shape progressively more ugly to make it fit the way the universe behaves; the way we know about. And there are more parts we have to tack on to this perfect shape, and then the less perfect the shape becomes, the more ugly it becomes. Then the more determined we are that underneath this perfect shape with the bits bolted on—that if we just open it up and unfold it or look at it from a different angle or reflect it through ninety degrees—well, that these imperfections vanish and we get back to something ideal again. So for me the universe isn't something concrete like that. It's an idealization I have in my mind. I can't put it any better than that.

Notable Quotes by Tara Shears

- We would like to think that the universe is simple and beautiful. By "beautiful" we do simply mean our fascination and by "simplicity" we mean an elegance in the sparseness of the description you need to describe so much of it.

- Our fundamental understanding of the universe is breathtaking in a very humble manner because there is no reason on earth that we should be able to understand the universe. And there is no reason on earth that we should be able to understand and devise a theory that describes so much of what we see experimentally.

- The universe evokes a feeling similar to when you look at great art or when you listen to great music that moves you. It takes away all your concerns from everyday life and it confronts you with the consciousness of the transience of it all.

- The main interesting questions in physics today all have to do with the universe. Are those elementary particles that we study really elementary or is there something beyond? What is the universe really like?

8 The Theorist: John Ellis

Jonathan Richard (John) Ellis was born on July 1, 1946, in London. After high school, he attended the University of Cambridge, where he was awarded a PhD in theoretical high-energy physics in 1971. In the following years, Ellis worked at the Stanford Linear Accelerator Center and at Caltech (California Institute of Technology) in the United States. Since 1973, Dr. Ellis has worked at CERN. Ellis is the James Clerk Maxwell Professor of Theoretical Physics at King's College London. Dr. Ellis' research interests are particle physics beyond the Standard Model, string theory, CP violation, the Higgs boson, and related fields of high-energy astrophysics and cosmology.

Jonathan Richard Ellis.
(© Michael Krause)

From 1988–1994, Ellis was Head of CERN's Theory Division; he is also a member of the CLIC (Compact Linear Collider) committee for the next generation of accelerators at CERN. In addition, Ellis is responsible for the connection of CERN to non-member states. Ellis was awarded the Maxwell Medal and the Paul Dirac Prize. He is a Fellow of the Royal Society and a Commander of the British Empire. John Ellis is the most cited theoretical physicist of all time, and he forged the term "Theory of Everything" (ToE).

Question: When you were a boy, and you looked at the sky, what was your feeling? What was nature like for you as a boy?

Ellis: When I was a kid, I lived on the edge of the countryside so I often used to walk across the fields. I was also interested—probably like all kids—in astronomy and I remember that I had a book where I had to tick off all the constellations that I could see with the naked eye. I think one of the things I learned to obey very quickly was not just observing but trying to understand the mechanisms behind what you could see. And I think that was what attracted me into science, and specifically into physics.

Question: So you were mathematically interested in it, how the constellations work?

Ellis: No. It is more a question of just being interested in how stars shone: what was the galaxy, what was the star? Rather in the particular geometric pattern which you can see with the naked eye, which actually is not very informative. Of course, it's important that you can see the Milky Way. It tells you that there is some structure in the clustering of stars but if you just look at the nearby stars and form the constellations that you are familiar with, they don't have any fundamental significance.

Question: So how did it go on then? How did you discover the universe?

Ellis: What I remember was, probably when I was twelve years old, something like that, I used to read a lot of books from the public library. Kids under the age of fourteen could not take out the adult books so the kids' books, the fiction books, were just not interesting. What was interesting was, of course, the non-fiction books. So I read quite a bit of history, quite a bit of science, various different subjects—at least the way that I remember it. I figured that physics was the most fundamental of all the sciences and so that was the thing that I got interested in at that stage. Perhaps it's also worth remembering that around that time was when Sputnik was launched. I specifically remember Sputnik as being an influence on the way I looked at the world.

Question: When did you become interested in Quantum Physics?

Ellis: I got interested in physics and also astrophysics and cosmology when I was around the age of twelve. Then, when I was around the age of fourteen, I had to make a choice whether I wanted to study more scientific subjects at school or more classical subjects and I decided that I wanted to study science, much to the disappointment of my headmaster who actually would have liked me to study classics. He made an attempt to get me to change when I was fifteen, but I said

"No, no!" I stuck to science. I think I was quite fortunate in having a very good mathematics teacher and a very good physics teacher at my high school. And at least the way that I remember it, I really decided I wanted to do something like theoretical physics and they certainly helped me along the way, but I don't think it was much to them that they persuaded me because that was what I wanted to do. I knew already that that was the sort of thing I would like—and they, as I said, very much helped me.

Question: Where do you get your power from to go on?

Ellis: I am really excited about anything that is new—some new discovery about the way that the universe works, and I am excited by new discoveries in astronomy, in new discoveries in paleontology, new discoveries in archaeology. But of course what really turns me on the most is new discoveries in physics, and this is for a combination of different reasons, I think. One reason is that it has explicative pay. We reach a new understanding of something in physics; this can help us understand the phenomena in the universe and perhaps explain some very fundamental aspects of the universe. It is also to say that often new discoveries in physics are very beautiful and that's also something that I find very satisfying.

Question: What's the beauty of it?

Ellis: The beauty is when there is a phenomenon which somehow seems complicated and difficult to understand but the thing you discover is there is actually a very simple reason why things work out that way. And perhaps also you can tie together different phenomena *a priori* completely disconnected. I think that one of my psychological urges is to seek connections or unified explanations of things and there is nothing I like more than to see some connection of two objects, two phenomena that seem *a priori* to be completely unrelated.

Question: So is nature easy to understand or is it very complicated? And what makes it easier to understand the complications of nature?

Ellis: Nature is certainly not easy to understand. The deeper you go inside the explanations of how the universe is, the way it is obviously, the more complicated it becomes. It's like digging a hole on a beach where you have to get rid of all the sand on the top if you want to get down to the rock underneath. But that's of course the attraction. If it was too easy then it wouldn't be so intellectually satisfying, shall we say.

Question: In the history of science, where are we now?

Ellis: We are at a really fascinating turning point—I would say—in fundamental physics. I have been a physicist for quite some time now, and the only previous occasion that I can remember things being as fluid and uncertain as they are now was back in the 1970s when new particles were being discovered and eventually that led to the general acceptance of what we now call the "Standard Model." So that describes particle phenomena in the laboratory very well; a number of questions it leaves open, on the cosmological side, for example. The Standard Model cannot explain dark matter, for example.

The Standard Model

The so-called Standard Model of particle physics summarizes today's knowledge about the fundamental particles. The model describes all known phenomena of the microcosm in a classification similar to the periodic table of elements.

The Standard Model allows for unification of the electromagnetic and weak nuclear force. It precisely describes almost all particle reactions observed to date. The model is based on quantum field theory. The fundamental objects, or particles, are considered as fields in space-time (field theory) that can only be changed in certain packets or quanta. The Standard Model obeys the laws of the special theory of relativity, i.e. it is "relativistic."

The Standard Model includes all 61 particles of which matter is created from, and all the interactions between them, which are transmitted by force particles or bosons. There are a total of 12 matter particles, 6 quarks and 6 leptons, and their antiparticles. Quarks and leptons consist of three families or generations. All known matter is composed of these particles; they are the building blocks of the universe. Vector bosons (photons, gluons, W and Z bosons) mediate the interactions between the particles but they can also be particles, such as the photon, which is also the quantum of electromagnetic waves.

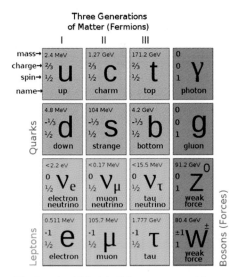

The Standard Model of elementary particles: 12 fundamental fermions and 4 fundamental bosons (without the Higgs boson).
(© Wikipedia, Fermilab, DoE)

The Higgs mechanism explains how the elementary particles get their mass and why the force carriers of the weak interaction, or bosons, have mass while photons are massless. Scientists believe that there is a Higgs field everywhere in the universe, and by interacting with this field the particles get their mass. To prove the existence of this field, the observation of the Higgs boson is necessary as well as the definition of all its characteristics. The discovery of a "Higgs-like" particle was announced by CERN on July 4, 2012, with a certainty of 99.99999%, or 5 Sigma. The discovery of the Higgs boson completes the Standard Model with all of its required parameters.

Although the Standard Model of particle physics can explain almost all observations made so far, it is incomplete; it has too many properties that can only be determined experimentally, such as mass, spin, etc.

Open questions regarding the Standard Model:

- What is gravity? The Standard Model does not include gravity or gravitational interaction.
- Why is gravity so much (10^{32}) weaker than the electroweak force (hierarchy problem)?
- Does the Higgs boson really exist (this question has now already been solved), and does it really have all the properties required by the Standard Model?
- If the Higgs boson gives mass to matter, why do particles have different masses?
- Why are there no significant remains of antimatter in our universe? According to the Standard Model, matter and antimatter should be equally distributed in the universe.
- Why are there exactly three particle generations or quark-lepton families?

The Standard Model is indeed the basis of modern particle physics, but there are numerous efforts to expand or replace it. There are many alternative models for physics beyond the Standard Model; they can be found in the field of "New Physics." One of the best known models is the "Grand Unified Theory," or "GUT." This new theory is aimed at unifying the three existent basic forces into one single force. Supersymmetry is another model; this theory postulates a symmetry between bosons and fermions. Every particle of the Standard Model has a symmetric partner particle in this model. Until today, however, there has been no evidence found for the existence of supersymmetric particles. Other approaches towards an extension of the Standard Model are the theories of quantum gravity or string theories.

Ellis: The LHC has the capacity to take us beyond the Standard Model and perhaps explain some of the mysteries, such as the origin of dark matter. Certainly that is one of the things that I work on a lot of the time. The advantage of the LHC is that it is the first accelerator to explore directly a whole new range of energies—lots of magnitude, perhaps more than we have been able to explore in the laboratory previously. We got plenty of good reasons for thinking that there is some new physics out there like dark matter, like the Higgs boson, but we don't know. We just don't know what it is and that's what makes the present moment so fascinating.

Question: What will be the cornerstones of New Physics?

Ellis: I think one of the cornerstones surely is the Higgs boson. But I think it would be too heuristic to think that the idea of an elementary Higgs boson as proposed back in 1964 is the whole answer. Maybe it's not even part of the answer. Maybe there is a completely different answer. But whatever the answer is, it has to emerge in the experiments now taking place at the LHC. So that would be one cornerstone. I think another cornerstone is dark matter. In many theories of dark matter, it is made up out of particles which once upon a time were sitting interacting inside the primordial cosmic soup. In those theories, that dark matter particle would have to weigh somewhere between 100 GeV or 100 times the proton mass, and 1,000 times, so that again puts it within reach of the LHC. So I think that's potentially the second cornerstone of new physics.

Question: What about dark energy then? More than 60% of the universe is made up of dark energy. Can we tap into that?

Ellis: Dark energy is a very, very big puzzle. The naïve theories that we write down—think for example about the Higgs theory or supersymmetry—all those theories predict the existence of dark energy. So you might say, "Ah, that's good," because of a successful prediction. Well, actually, no, because it tends to predict many, many, many, many, many orders of magnitude—too much dark energy. So the big puzzle is not so much why is there dark energy? The puzzle is more, why is there so little dark energy? Now for a while, theorists tried to prove the hypothesis that there is no dark energy at all and I think for many of us it came as a bit of a shock in the 1990s when the astronomers and the cosmologists and the astrophysicists came back to us and said, "No, no, there is dark energy. It's just that it's very small." And quite frankly, I don't think we have a very good idea how to explain that small amount of dark energy. Now my point of view on that is that

Halo of dark energy surrounding our Milky Way galaxy. (Artist's impression, © ESO/L. Calcada)

the best way forward is to study the fundamental physics which should contribute to dark energy and which, according to our calculations, appears to give way too much dark energy. And if we study that maybe we can find out what we're doing wrong and maybe that will then enable us to come up with a better theory which gives you a smaller amount of dark energy, perhaps with a new experiment.

Question: Let's go back to your own story. What are the properties of a physicist?

Ellis: I think it is clear that you really have to be obsessed by understanding more about the way the universe works. You have to be convinced that this is an important thing for at least a few people on the planet to be doing. There is something like seven billion people on this planet, and something like one in a million of those people is actually working in particle physics. I think that is not unreasonable that of all the people that are on the planet, a few people, one in a million, should share this obsession. But it really must be an obsession. It's not something which you do as a hobby. One of the things that really drive you to get up when you wake up each day, must be this urge to understand the universe better. If you're not committed to it, if that's not one of the most important things in your life, forget it.

Question: In your life, was it ever satisfying to wake up with that urge and find something maybe very different or nothing? Did you have those fantastic moments?

Ellis: There are various levels which you get award or satisfaction from feeling that you understand things better than you did yesterday. If you're very, very lucky, sometimes you may actually have the impression that you are the only person on the planet who understands that particular thing. Most of the time though it's more a question of making some sort of incremental progress in whatever research project you're doing—little progress, if you like. Every once in a while there is a big, "Eureka!" And, well, I have been lucky enough to have had one or two big "Eurekas" in my time but usually it is just a small eureka!

Question: What were those one or two "Eurekas"? How did it feel?

Ellis: I remember that perhaps my most satisfying eureka moment was when I was walking down a corridor here at CERN and just around the corner I realized how one could discover the gluon particle. This was back in the mid-1970s, when we had a generally accepted theory of the strong interactions—called QCD—and according to the theory, there should be particles holding quarks together, called gluons, in the same way that photons hold electrons together with nuclei inside atoms. And so pretty much everybody, at least all theorists, were convinced that these gluons existed, but there was no direct experimental proof. And I was just walking back to the cafeteria one afternoon and I'd been having a discussion about scattering particles at high energies and I suddenly had this idea that that's the way in which we could really discover the gluon in a very clean way. Not in scattering particles, but in annihilating electrons and positrons, and then having particles coming out. And then every once in a while, one of those particles would be a gluon and you could see it very clearly and distinctly because you weren't colliding garbage, because stuff was coming out of very simple collisions. So, with a couple of colleagues, we calculated this process, we wrote up a paper, talked to the experimentalists. A couple of years later, they did the experiment and they discovered the gluon.

Question: Do you remember your feeling in that moment?

Ellis: It is funny, sometimes ideas apparently come out of the blue, with no obvious direct stimulus. And so there must be some subconscious level, some mechanisms within our brains where different ideas are swirling around and then sometimes they bump into each other and then perhaps they create something new. And so that was what happened on that particular occasion. I was lucky that the swirling around happened to produce this particular idea.

Question: How was it when you talked to your colleagues about that idea?

Ellis: That was not a particularly complicated idea to explain, but I should say that afterwards when we tried to explain it to other people, often we had difficulties. I remember beginning a seminar about it at the laboratory in Germany, where the discovery was later made, and at least some of the theorists there were very skeptical and giving me a hard time. OK. I didn't care particularly. I think that showed that they didn't understand the way the theory should work.

Question: What was the most exciting moment for you in physics? What was the biggest challenge?

Ellis: Well, there have been other challenges and there have been other exciting moments. I mentioned the 1970s were the previous most exciting period. It was a time in 1974 when new particles were being discovered almost every week and it was very exciting trying to figure out what these new particles meant: what they were; what was the organizing principle? So that was a very exciting period. In the 1980s the W and Z particles were discovered here at CERN and I remember actually during that academic year I was in California. I remember that I was at a conference in California I organized for one of my colleagues to come out and present the latest results which actually were the first W particles so that was pretty exciting, seeing these computer printouts and … yes, these must be the W particles, at long last. I expect something similar to happen soon with the LHC.

Question: What do you expect?

Ellis: Well, there is the Higgs boson. I'm optimistic about finding dark matter particles. As far as the Higgs is concerned, we knew that it was not going to be easy to discover it at the LHC and I think it's remarkable how quickly the accelerator and the detectors have got working. Dark matter; I am maybe just a little bit disappointed that we haven't seen any signs of it yet. I and others made estimates of what the masses of these dark matter particles might be. In simple models, those don't seem to pan out but we're still optimistic and we're looking forward to the LHC data.

Question: Where in the history of science would the LHC be?

Ellis: I think that the LHC marks a turning point which will potentially enable us to turn a new page in fundamental physics. If you look back over the 20th century, I think it was a golden century for fundamental physics, with many very important discoveries and new principles of nature revealed. In particular we saw emerge several new layers of what you might describe as the "cosmic onion," so if you think back to the end of the 19th century, I think physicists finally were prepared to accept the existence of atoms. Almost, but not completely. It was just before the end of the 19th century that the electron was discovered. People realized that actually atoms were not atoms in the sense of being indivisible objects but actually they had things inside them—a complicated structure. So the first new layer of the onion was to peel off the electrons on the outside from the nucleus in the inside of the atom. The next step was to open up the nucleus and discover that it was made up of protons and neutrons. The next layer to be peeled off was protons, and neutrons are complicated objects; they contain quarks. All this was done during the 20th century. I believe that we are probably about to peel open another layer of the onion. I don't know what that is going to be. Maybe we are going to discover that the Higgs boson is a composite object made up of other things or maybe we are going to discover that the Higgs boson is just one of an enormous family of new particles. I don't know.

Question: What would this new "particle family" look like?

Ellis: There are different theories about this. I think most theories predict that the Higgs boson is not alone. Most theories about physics beyond the Standard Model suggest that it is part of a sort of family of new particles so I think the minimal extension, if you like, of the Higgs idea is supersymmetry. Supersymmetry postulates that for every known particle there is some other particle which differs in the internal property of spin. All the particles are spinning around; some of them spin at different rates from others. Supersymmetry predicts that for every known particle spinning at a certain rate, there would be another particle with identical, for example, electric charge, but they would be spinning a little differently. According to that theory, the number of elementary particles should be doubled. Now that is the most, in some sense, economical theory in terms of new particles.

Then there is the idea that the Higgs boson might be composite. If it is composite—made out of some sort of smaller things—then you can combine those smaller things in many different ways. In fact, potentially, an infinite number of ways, in the same way that you can combine quarks and gluons into an enormous number of so-called "elementary particles." Like originally we thought, "Well, there's just the proton and the neutron." And then people discovered the pion, they discovered the kaon, they discovered literally hundreds, if not thousands, of other particles made up of quarks. It could very well be that when we open up the Higgs we'll discover there's something inside and those things inside could be combined together in hundreds, if not thousands, of different ways to make a whole spectrum of new particles.

Columbus Landing in the New World on October 12, 1492, engraving by Theodore de Bry, 1594 (Netherlandish, 1528–1598). (© US Library of Congress)

Question: Columbus set off to find the sea passage to India, but instead he discovered America. Could this kind of analogy also be true for the LHC?

Ellis: I like that analogy. In fact, I used it a couple of years ago in an article I wrote. As you say, Columbus set off to find Asia, and he discovered America. So it could very well be that we are going to start up the LHC to discover the Higgs boson, and instead we are going to discover…I guess by definition we don't know what we are going to discover, right?

Question: Is this kind of uncertainty the thing that satisfies you, personally?

Ellis: Yes, it is very exciting. It has often been said that if we knew what we were doing it wouldn't be research. That is certainly the way I feel about the LHC. LHC physics is much more exciting because we don't know what we are going to find—although it would be a tremendous achievement for theoretical physics, experimental physics, accelerator physics if we do discover the Higgs boson, it would actually be more exciting if we'd discovered something more complicated.

Question: What would be your craziest theoretical idea?

Ellis: I think one of the biggest open problems in fundamental physics is how you combine quantum mechanics and general relativity so here we are about a century after quantum mechanics was stated to be discovered and almost a century since general relativity was discovered by Einstein. And you can regard it as a terrible failing of theoretical physicists that we haven't yet succeeded in combining the two. Of course, we have ideas about how it might be done; string theory is the foremost among of them, but we have no real experimental probe of such theories at the moment. And this is why our string theorists often get frustrated, why people often criticize string theory because they say, "How do you test it?" Well, to test any quantum theory of gravity is very, very tough.

That is why it is a very interesting subject to think about. Now, if it was easy, it would have been done. I think you have to think outside the box if you want to imagine how you might consider testing a quantum theory of gravity and the way that I would approach it is to say, "If you want to combine quantum theory with relativity, very likely either one, or the other, or both has to be modified." Now, as soon as you say that, you are immediately a heretic. Immediately about 90% of your audience will sort of shake their heads and say, "Ellis has finally gone nuts." Right, OK, that's fine. I am old enough that I can go nuts. So if either relativity or quantum mechanics is broken, how might they break? And how could you test that? So one of the themes in my research—every few months, maybe every year or so, I write another paper about that—is trying to figure out new ways of looking for the breaking of relativity or quantum physics, and also analyzing [based] on the constraints of such a radical idea.

Question: How could that—theoretically—be done?

John Ellis—How could relativity or quantum mechanics be modified? (© Michael Krause)

Ellis: Perhaps to say that a theory is broke or that a theory needs mending—perhaps that is not the right way of describing it. I mean, relativity clearly works over a very large range of energy scales and distance scales. Quantum mechanics certainly works. So I think it is more a question of pushing the envelope of where those theories apply, and perhaps finding around the edges there is some small deviation which we have to take into account. So I think the analogy potentially is with Newton's theory of gravity.

Newton's theory of gravity works just fine for all the objects in the Solar System, as long as you do not look at them too carefully. If you look very, very carefully at the orbit of, say Mercury, then you discover that there is a small precession of its orbit which you can't understand with Newtonian gravity. But most of the time Newtonian gravity works just fine. And so I like to think, or say that it might be true of relativity, and the same might be true of quantum mechanics; that for most purposes they work just fine but it is only when you really push the boundaries that you discover that there is something else which lies beyond and modifies them in some way.

Question: Is the theory of relativity falling apart or will it sustain itself?

Ellis: At the moment quantum mechanics and relativity work perfectly. There is absolutely no sign of any potential modification of either of them. At the moment, it is just speculation, idle speculation. But still, I think it is useful to think what might be out there beyond those theories.

Question: What do you think, what might be there?

Ellis: For example, we are told that massless particles always travel with the speed of light and the speed of light is a constant which is independent of the frequency of light or, if you prefer, the energy of the photon, of the quantum of light. Maybe that is not true. Maybe quanta with different energies actually travel at slightly different speeds. Now you might think that the most natural possibility would be that the more energetic electrons travel at higher speeds; that certainly happens with ordinary particles. Pump more energy into your car and it goes faster. But maybe it is around the other way with massless particles. This is an idea that we proposed about 14 years ago or something like that. And we have proposed various different experimental tests. Various different observations have been made. They don't find any effect of that type. I think those people think that it's a completely crazy idea, but still.

Notable Quotes by John Ellis

- There is nothing I like more than to see some connection of two objects, two phenomena that seem *a priori* to be completely unrelated.
- I think it would be too heuristic to think that the idea of an elementary Higgs boson as proposed back in 1964 is the whole answer. Maybe it's not even part of the answer. Maybe there is a completely different answer.
- The big puzzle is not so much why is there dark energy? The puzzle is more, why is there so little dark energy?
- There must be some subconscious level, some mechanisms within our brains where different ideas are swirling around and then sometimes they bump into each other and then perhaps they create something new.

9 Oersted — Ampère — Faraday — Maxwell

Hans Christian Oersted (1777–1851)

In 1820, the Danish physicist and chemist Hans Christian Oersted discovered the magnetic effect of electric current by connecting the two poles of a battery and holding a compass to the wire. When the power was switched on, the magnetic compass needle no longer pointed north but instead at right angles to the battery wire. Oersted thought that there had to be a direct connection between the electricity and the magnetic force, but believed that gravity, electricity, and magnetism were all just different manifestations of one single and completely unknown force.

André-Marie Ampère (1775–1836)

The French mathematician and physicist André-Marie Ampère repeated Oersted's experiments. Ampère constructed a new hypothesis that the electric current generated a magnetic field within the wire, i.e., the electric current had to be the cause of magnetism. Ampère determined the mathematics between current and magnetic field. The direction and force of this field depended on the direction of motion of the current and the shape of the wire. Ampère described the field forces by the right-hand rule: A) point thumb in direction of current and B) the fingers will curl in the direction of the magnetic field. This concept of lines of force was later adopted and generalized by Michael Faraday.

Michael Faraday (1791–1867)

> "Nothing is too wonderful to be true, if it be consistent with the laws of nature, and in such things as these, experiment is the best test of such consistency."
>
> — Michael Faraday's Diary, March 19, 1849

Michael Faraday was trained as a bookbinder and his scientific education was more or less self-taught. Nevertheless, he became one of the most important scientists of all time and one of the most prolific; he conducted about 30,000 experiments and published more than 450 scientific articles and books.

During his apprenticeship as a bookbinder, Faraday developed an interest in chemistry. In 1813, Faraday became the assistant of Sir Humphry Davy (1778–1829), professor of chemistry at the venerable Royal Institution in London. Davy, a scientific pioneer and the inventor of the arc lamp, promoted young Faraday as much as possible. During an 18-month tour through Europe, Faraday met, among others, André-Marie Ampère in Paris and Alessandro Volta in Italy. After this trip, Davy let Faraday set up and prepare experiments. Faraday soon conducted his own experiments, held his own lectures, founded private research circles, and offered a wide range of service analyses. In 1820 he was known as the best chemical analyst in the United Kingdom.

Faraday made his greatest contributions to science in the field of electrical engineering. In 1821, shortly after Oersted had discovered the phenomenon of electromagnetism, Faraday built an apparatus to produce what he called "electro-magnetic rotation": a continuous circular motion from the circular magnetic force around a wire. His experiment proved that electricity is able to do work. Faraday thus laid the foundations for the development of generators, dynamos, and electric motors ("Historical Statement Respecting Electro-Magnetic Rotation," 1823).

Faraday also made great discoveries in chemistry and optics. In 1823, he succeeded in liquefying several gases (chlorine dioxide), which proved that the states of solid, liquid, and gas can be converted into each other. In 1825, Faraday discovered benzene, an aromatic hydrocarbon and basic building block of organic chemistry. In 1827, Faraday published his further studies and discoveries in chemistry, titled "Chemical Manipulation." Over the years, Faraday became the most respected scientist of his time. In 1824, he was elected as a member of the Royal Society, and in 1825 he succeeded his mentor, Sir Humphry Davy, as director of the laboratory at the Royal Institution. In the following years, Faraday's popularity grew steadily as he opened the Royal Institution for public lectures. In his famous Christmas lectures, he presented science to the general public to inspire—and generate revenue for the institution.

Electromagnetic Induction—Classical Field Theory

In 1831, Faraday returned to his studies of electricity and made what is arguably his most important discovery, electromagnetic induction, using an apparatus known today as a transformer. He explained his results with the presence of magnetic force lines. Faraday thus created the basis for the electromagnetic field theory: electricity was generated within a coil only if the magnetic lines of force in a moving magnetic field intersected each other.

In further experiments, Faraday tested the properties of electricity from various sources (e.g., Volta battery, static electricity, etc.), concluding "that electricity, whatever may be its source, is identical in its nature" (*Philosophical Transactions*, 1833). In 1832, Faraday began his experiments in electrolysis, at the time called "electrochemical decomposition." Electrolysis was economically interesting; for the first time, chemically pure metals could be produced. In his publications, Faraday coined the terms ion, electrolyte, and anode and cathode for the two electrodes, and he formulated in his report ("Researches in Electricity," 1834) to the Royal Society the laws of electrolysis:

- 1) The mass of the substance liberated or deposited on an electrode during electrolysis is directly proportional to the quantity of its electric charge.
- 2) When the same amount of electricity is passed through different electrolytes, the amount of different substances deposited or liberated are directly proportional to the equivalent weight of the substances (law of equivalent proportions).

Faraday's laws describe the relationship between the amounts of substance deposited on the electrodes and the flow of currents. Faraday's second law represents an important connection between matter and electrical charge; it could only mean that both matter and electricity are governed by an underlying atomistic structure—there had to be diverse constituents of matter (e.g., atoms, ions, or electrons). However, the next steps in the development of the atomic theory—based on Faraday's laws—were made much later by Hermann von Helmholtz (1821–1894, "Imperial Chancellor of Physics") and Robert Millikan (1868–1953, 1923 Nobel laureate in Physics).

Michael Faraday was convinced that all the forces of nature are connected by a superior law, summarizing gravity, light, electricity, magnetism, heat, and even those forces in the matter itself. This unification of all forces is still a dream of physicists. Faraday himself was a deeply religious man and a member of the Christian sect of Sandemanians. Perhaps this belief also greatly affected his desire to summarize everything known (and unknown) under one single law—the ultimate law of God.

Electromagnetism: James Clerk Maxwell

"Faraday, in his mind's eye, saw lines of force traversing all space."
— James Clerk Maxwell

James Clerk Maxwell (1831–1879) was a brilliant Scottish physicist. In 1841, being just ten years old, he won the mathematics medal of the Edinburgh Academy. At the age of 16, Maxwell left the Edinburgh Academy to enroll himself at the University of Edinburgh. In 1850, he enrolled himself at the University of Cambridge and, before the end of his term, transferred to Trinity College. In 1854, he graduated from Trinity College with a degree in mathematics. In 1856, being just 25 years of age, Maxwell was offered a professorship at Marischal College in Aberdeen, which he accepted. In 1861, he headed to London, where he was granted the Chair of Natural Philosophy at King's College. In 1871, Maxwell was elected as the first Cavendish Professor of Physics at Cambridge.

Maxwell admired Faraday's experimental observations and began to transfer them into mathematical formulas. His first major essay, "On Faraday's Lines of Force," was published in 1856; here he described Faraday's lines of force as imaginary tubes containing an incompressible fluid. In 1864, Maxwell published "A Dynamical Theory of the Electromagnetic Field," a mathematical theory of electromagnetic fields, with the Royal Society, London. In this historical presentation, four of Maxwell's equations (of originally 20) were presented for the first time. With these formulas, all electromagnetic phenomena found by Faraday could be explained mathematically. Maxwell transformed Faraday's ingenious experiments into the mathematically rigorous form of the field physics. "Maxwell's wonderful equations" became the foundations of modern electrodynamics. They include the induction law and the laws of electromagnetic waves.

In addition to his research regarding electrodynamics, Maxwell made numerous scientific discoveries in other areas. In 1855, he formulated his own color theory and explored color blindness. In 1861, he made the first color photograph in the world. In 1857, Maxwell observed the rings of Saturn and claimed that the rings were composed of individual rocks. It took one hundred years before NASA's Voyager spacecraft was able to confirm Maxwell's prediction.

1) Maxwell equation ("Ampère's law") describes Ampère's circuital law. Each time-varying magnetic field generates an electric vortex field.
2) Maxwell's equation ("Faraday's law") describes Faraday's law of electromagnetic induction. An electromotance is produced in a circuit when the magnetic flux through the circuit changes.
3) Maxwell's equation (electrical source): Electric charges are the sources of electric fields.
4) Maxwell's equation (magnetic "source"): The field of the magnetic flux is source free. Magnetic fields always induce magnetic vortex fields.

Maxwell went on to study his equations and obtained a speed of electromagnetic waves, which corresponded to the speed of light. Maxwell's comment was:

> This velocity is so nearly that of light, that it seems we have strong reason to conclude that light itself (including radiant heat, and other radiations if any) is an electromagnetic disturbance in the form of waves propagated through the electromagnetic field according to electromagnetic laws.
>
> — "A Dynamical Theory of the Electromagnetic Field," *Philosophical Transactions of the Royal Society of London*, 1865

The full meaning of this sentence—light is nothing different than an electromagnetic wave—could not yet be proven at that time. In fact, it would take decades until Maxwell's equations were completely understood. Maxwell himself was convinced that light required a medium to be propelled in, the so-called "ether." In 1878, he wrote about it in the *Encyclopaedia Britannica*:

> There can be no doubt that the interplanetary and interstellar spaces are not empty, but are occupied by a material substance or body, which is certainly the largest, and probably the most uniform body of which we have any knowledge.
>
> — *Encyclopaedia Britannica, Ninth Edition*, 8: 568–572

Maxwell's electrodynamics combined all previous observations, experiments, and equations of electricity, magnetism, and optics together into one theory. Maxwell's equations showed that light, electricity, and magnetism are all single manifestations of one phenomenon: the electromagnetic field. Maxwell's electrodynamics, together with Newtonian mechanics, is the basic framework of classical physics. Maxwell's equations are considered the second great unification in physics after Newton. According to Richard Feynman, they are "the most significant event of the 19th century."

> *"Was it a God who wrote these lines …?"*
> — Boltzmann on Maxwell's equations (quoting from Goethe)

10 The Communicator: Rolf Landua

Rolf Landua: What is the universe, and how is it relevant to us? (© Michael Krause)

Rolf Landua grew up in Wiesbaden, Germany. He studied physics at the University of Mainz and graduated in 1980 with a thesis on "exotic atoms." In the same year, Dr. Landua went to work at CERN, becoming a CERN fellow from 1982 to 1985. Landua focused on antimatter research within the LEAR (Low Energy Antiproton Ring) experiment (dealing with antinucleon-nucleon reactions, meson spectroscopy, and the search for exotic quarks and gluons). In 1996, anti-hydrogen atoms were observed for the first time within the LEAR experiment. This CERN coup had great media coverage, and was even a model for a Hollywood blockbuster film ("Angels and Demons," directed by Ron Howard, 2009). Dr. Landua is cofounder of the ATHENA experiment (AnTi-HydrogEN Apparatus). He was the ATHENA spokesman from 1999 to 2004. In this experiment, millions of antimatter atoms were produced for the first time.

Today, Rolf Landua is committed to the intensification of teaching science at schools. In 2005, he became head of the Education and Public Outreach Group (PH-EDU) at CERN. Landua's group organizes national and international programs for physics teachers, with approximately 1,000 participants per year. The aim of the program is to make modern physics research in European classrooms more readily available. His group is also responsible for the exhibitions and visitor programs at CERN. Rolf Landua is the author of popular books about CERN, such as *On the Edge of Dimensions*. He received the 2003 Award for Communication from the European Physical Society. As useful ingredients for a professional career at CERN, Dr. Landua recommends patience, enthusiasm, a good knowledge of English, and friendly interaction with foreigners.

Question: In your school days, what were the things you were interested in?

Landua: I loved science fiction; it was my favorite. When I went home, that made me think about the universe, particles—everything.

Question: What kind of science fiction, something like "Electrical Experimenter"?

Landua: Yes, also every kind of science fiction I could get a hold of, and I really loved to dream about things and to find out what there could be. You know, what the possibilities were.

Question: What were the key topics then?

Landua: If you look into science fiction of those days you will find everything we are discussing today. You will find extra dimensions; you will find spacecraft that will go faster than light. You will find the questions: why is the universe as it is? Are there wormholes? About black holes, it was all there already. Today, science has caught up a little bit with fiction.

Question: At that time, when you looked at the sky, what did you imagine?

Landua: I saw a promise. I saw something which was bigger than my environment, which was sort of a promise, a temptation to explore the space—to find out what I am made of. It was just a dream, like some people dream to climb on a mountain or to explore the oceans. It was a dream to explore space, to explore the universe. This was what it meant for me.

Question: When did you decide to make your dream a profession?

Landua: Well, there were several steps. It started off as a 15-year-old. I found chemistry very important because chemistry had all these things about atoms and molecules. That's what everything is made of but then I noticed that the part in chemistry that really interested me was physics. And so a few years later I decided that I wanted to study physics to find out what everything is made of and where it comes from. I studied physics and then of course, without surprise, I ended up in particle physics because that is really the place where you think about the most fundamental particles and laws of nature and so on. I tried to get into university and the crew that works at CERN. And so, finally, I ended up as a postdoc in an external group at CERN and then I managed to get a hold on a post at CERN. Now, I am a physicist here at CERN.

Question: How did your family react when you told them that you were going to be a physicist?

Landua: They looked at me and said, "What does a physicist do?" I tried to explain but it didn't go so well because my father knew physicists working at Hoechst, a big chemical company. They were really a minority, and they measured the thickness of foils over a radioactive source. My father said, "Is that what you want to do? Measuring thickness of foil?" And I said, "Well, physicists do more than just that, you know?" But still, I think they really didn't get it. After a while they understood that it was quite an interesting field and that I was very happy doing it and so they were happy, too.

Question: What are your hobbies?

Landua: I love sports, reading, holidays, traveling around—these are my hobbies. But if you have three kids then you don't have so much time to read. I am coming back to my hobbies now. Maybe music—but I am not very good at it.

Question: What kind of music?

Landua: Oh, with notes. I play a little bit of piano. I played much better but now I am looking forward to the next years, to learn a little bit more.

Question: Is there any connection to physics?

Landua: Yes, if you would strain the connection, yes, of course. Of course, it's acoustics, and certain harmonies, but otherwise, no.

Question: Is there some connection to physics in your private life?

Landua: I hope not. It is usually a recipe for disaster if you bring in physics into your private life.

Question: You are teaching teachers. What kind of special program is this?

Landua: I stem from 20 years of research in antimatter, and I noticed that every time when I was invited to give a talk to school children coming to visit CERN, how much they were really mesmerized by hearing about antimatter and everything that goes with it: the Big Bang, the transformation of energy into mass, and so on. I noticed that we are failing, I think, in schools to mesmerize children with science, to really make them excited about it. Science is presented as a dead subject, as a bunch of formulas which you have to learn in order to pass the exam, and the kids do not know what to do with these formulas and what they have to do with real life, and with this environment. At one point I decided maybe now it's time to help to get young kids more interested in science again because we have that problem not only in Germany but in many other countries.

Everybody goes into finance or history, or geography, social science, or whatever, but very little into physics or engineering and that has to do partially also with the fact that it is not very enticing, what they are doing. So I decided that CERN should make a bigger effort in order to carry over this excitement of scientists on to young people. And of course, in between us and the young people, there are the teachers. The teachers are the role model; they are the most important part of society, and very often that's not really taken care of in the right way. So I decided that we should have a group here which deals with teachers and sort of put them into a position, talk about modern science in an exciting and really appealing way so that kids can decide at the age of 16 that physics is really interesting—like I did because of the science fiction—and maybe to go into it and continue, and then go on and maybe become scientists but still understand what science is for, and how it works and why scientists say yes, and maybe no. I mean, just the whole process, the scientific method. For us, this is really a very successful program. We had over 5,000 teachers here in the last five years and they really like it. And I think it makes an impact even though it's just a drop in the ocean, of course, but it can grow, and I hope it will.

Question: What is so exciting and appealing about science?

Landua: Einstein once said something in the sense that the least understandable thing about the universe is that it is understandable. It's a fact that you can ask questions about the origin of everything, the foundation of everything, the basic building blocks. And if you do it properly, and if you follow a certain method, you can find answers. And you can test the answers and you can test if you have understood it. And you can understand it. The universe is understandable. And that's a very exciting thing because it has lots of applications. I mean that is sometimes not so well known—that you start with a simple question, like, what is the atom made of? Or what is electricity? Then you find out laws of nature which deal with electricity and magnetism, and so on. And suddenly you understand something and suddenly you can use this theory for your daily life. You would never have the light bulb if you haven't done experiments with electricity. Any type of research which you have put into improving the candles, even in a hundred years, you would have never found the light bulb.

Fundamental research is something to push frontiers, to go beyond what you know, and it is really exciting to ask that. It is natural curiosity that everybody somehow carries, at least as a child, inside. Sometimes it gets lost, and sometimes even with 50 or so, you are still curious. And that's a good thing as a researcher, to be curious.

Question: A good researcher has to stay curious like a child?

Landua: I think at any age curiosity is a good thing. It should be maybe paired with patience and with certain skepticism. Challenge old theories, always thinking, "Is it really true what I believe is true? Can I explain that with this theory?" So, curiosity is a good thing but there are a few other things that a good researcher should have.

Question: What are the things a researcher has to have besides curiosity?

Landua: Well, as I said, a good researcher has to be, first of all, skeptical. Skeptical, in a sense, that old theories might be wrong even though he believed in them for 30 years. Experiments may be wrong, and he should always challenge authorities because I think there's nothing worse than just to believe in somebody because he is old and famous. Patience is a really important issue because nothing in science is a quick thing. It doesn't come like that. It often takes years; here at CERN it takes tens of years until we make a step forward. All simple experiments, they have been done many years ago and now we are challenging nature at a scale which is much bigger than it was in the past so we have to really make an effort.

Question: Please explain, understandable for a layman, what are the people at CERN doing?

Landua: They are doing many different things but they all work for one common goal and this common goal is to find out how our universe looked like at a very, very tiny fraction of a second after the Big Bang—but only on a tiny scale. And for that they have to collide two particles, or protons, at the highest possible energy that we can achieve right now. In these collisions we recreate conditions as they were just a tiny fraction, a picosecond, after the Big Bang. Then new phenomena occur; we are looking into these phenomena and we try to disentangle what happened, and how did it shape our universe as it is today. That is the goal. But of course in order to achieve the goal, we have to build our tools. The tools are called "accelerators" in order to make these particles go as fast as the speed of light or almost the speed of light—and make them collide. Then we have to have detectors, which are like huge cameras; they are 25 meters in diameter and 40 meters long. These cameras can take snapshots of every collision and look at what happened, and they can look for the signatures of things we don't know yet. So that's basically what we do. Then you can break down in many divisions and professions what happens here at CERN. There are physicists, there are engineers, and there are technicians, administrators, the IT specialists, the electrical and mechanical and what other engineers. Here you find many types of different professions and they all work together for this common goal.

Question: If CERN was an organism, who would be the brain?

Landua: I think it is a little bit like in our brain. There are many cells in your brain but a cell is not a brain. And I would compare the researcher rather with a cell than with a brain. Basically, we here at CERN have in some sense one big brain which consists of many, many researchers which somehow behave like collective intelligence. We have 10,000 scientists here working at CERN—fortunately not working here all the time, otherwise the cafeteria would be awfully full. Say something like three or four thousand are here at a given time and they work on the analysis or the construction or the improvement and many other things, in permanent communication with each other. Taking into account what the others have said, and criticizing, they make progress together. It is not just one single scientist who sits in his small office and thinks, and at the end of his thinking he comes out and tells everybody what the result is. That's not how it works. It is really collective intelligence.

Question: In your teaching, you use a painting by Gauguin. Why did you choose Gauguin?

Landua: There are two reasons. The first reason is that it is a nice picture. I like Gauguin, and he had good taste in order to go to the South Sea to spend his life there, nice weather. But the real reason is because it was made in 1897. 1897 was also the year when particle physics started because the electron was discovered by

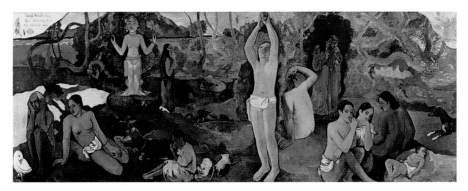

Where Do We Come From? What Are We? Where Are We Going? (Painting by French artist Paul Gauguin, 1897–1898.)

J.J. Thomson. It was the first discreet constituent of matter which ever was found; it was the birth of particle physics. And the other reason is the title of the picture which is, *Where Do We Come From? What Are We? Where Are We Going?* and these are basically the questions we are asking today at CERN.

What are we? That is the particle physics part.

What are the constituents of matter? Where do we come from? That's the question of the Big Bang, the origin of the universe, and the evolution from animated matter to the organization of matter on different levels to our existence.

And where do we go? That is the obvious question that you ask when you have understood all what happens in the past; then you want to know what is going to happen.

And of course the fourth question is what will we have for supper tonight? But that's Woody Allen who asked that question.

Question: Cosmology and particle physics, why have they appeared to merge?

Landua: They merged at a time when cosmologists discovered that in order to describe the first instants of the universe, the first few minutes of the universe, they needed particle physics. And it came also from particle physicists; they discovered that everything cosmologists sort of see in the universe has an immediate application in particle physics. I'll give an example. For a long time dark matter was known, but people did not know what particle may be responsible for it. And so, the cosmologists thought, "Well, it could be the neutrinos." But in order for neutrinos to be responsible for that, they would have to have a mass—something like ten electron volts or more in order to be relevant. Particle physicists could already say from their experiments that the neutrinos must be lighter than that so there was immediately a connection with the two. And that has stayed like that for a long time. If you want to describe the early universe, you need particle physics. You need to know what exactly existed at a given time in order to find out how to extrapolate from there to later times. So the two sciences have really come together.

Question: What would a world in "New Physics" look like?

Landua: I wish I knew. If I knew, I would probably soon be awarded the Nobel Prize. In the early days you had mythology. You had to make analogies between the life on Earth and how it could be on a big scale even though we did not know how big the universe was at all. The size of the universe is only known for about 70 or 80 years, after Hubble and his discovery. We believed for a while that we could understand the universe just by putting in enough of our type of matter—atoms, with protons, and neutrons, and electrons. Then we found out that the universe is made out of other stuff that we don't understand: dark matter, but also dark energy. I wish I knew what it is actually, because this is part of the reason why we do what we are doing with the LHC. The universe is made up out of 96% of stuff that we do not know about. That's really exciting! Because in some sense that provides us with a very good motivation to continue our studies, and I don't know really what will come out of this exploration because dark energy seems to be such a mysterious thing. The fact that suddenly, nine billion years after its start, the universe suddenly reaccelerates and expands faster, that's just absolutely amazing.

Edwin Powell Hubble (1889–1953)

> *"Equipped with his five senses, man explores the universe around him and calls the adventure Science."*
> — *The Nature of Science*, 1954

In the early 1920s, the American astronomer Edwin Hubble discovered—together with his colleague, Milton Humason (1891–1972), at the Mount Wilson Observatory in Los Angeles—that all spiral nebulae (galaxies) were living outside of our own Milky Way. With this discovery, Hubble became the pioneer of modern extragalactic astronomy. During further research, in 1929, Hubble discovered that the redshifts of distant galaxies increased proportionally with their distance. This meant that almost all galaxies move away from us, and the farther away they are, the faster they move away. Hubble's discovery, the expansion of the universe, represented a paradigm shift in cosmology. Hubble himself did not come to this conclusion, but labeled his observations with the term "apparent motion." The truth is the universe is not static, but expanding. It is arguably the most important cosmological discovery ever made.

Described by Georges Lemaître (1894–1966) for the first time, the "Hubble constant," or "H" in modern physics, describes the increase in speed of galaxies moving away from us. According to the latest measurements, it is about 74.2 kms per second per mega-parsec (= 3.08 million light-years).

In honor of Edwin Hubble, the Space Telescope "Hubble" bears his name.

> *"The history of astronomy is a history of receding horizons."*
> — Edwin P. Hubble, *Realm of the Nebulae*, 1936

Question: Is this not frustrating? More and more knowledge has been accumulated, and now we realize, "Well, there might be something else out there"?

Landua: I think it's not frustrating really. I think it's exciting. It's really something that makes you want to think more, make new hypotheses, test them in the lab if you can. It's really something which motivates you in my view. I think the most frustrating thing is if you are stuck somehow. If everything is there, you have a theory which describes everything and you have no idea where this theory comes from, it's like having a recipe for a good meal and somebody has written this recipe, and it works but you wonder, "Well, why this ingredient, why that ingredient, not

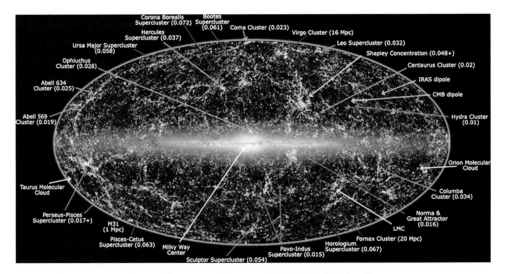

Large Scale Structure in the Local Universe. (T. Jarrett, NASA/IPAC/Caltech)

Rolf Landua: We want to understand what the universe is. (© Michael Krause)

double of that, and half of this?" So you want to understand that recipe of the universe and it is really a very exciting thing for curious people.

Question: Of these three items, which is the best representation of the universe?

Landua: Actually, none of the three. For me, the universe is something which really doesn't have a boundary. That is the biggest problem with these three things. For me, the universe is not limited in space nor time. It is just something, if you wish. It is a sphere with additional dimensions attached. We are inside a four-dimensional sphere, and we are living on the surface of this sphere, which is a three-dimensional object. At least that is my imagination, so you are not running into the problem of explaining what happens at the boundary because there is no boundary.

Question: Do you think that the universe is endless?

Landua: Yes, as a three-dimensional object it is endless. It may have a finite volume in the sense that even a three-dimensional sphere has a final volume but it doesn't have a beginning nor an end in the sense that a two-dimensional dog which crawls over the surface here can crawl as long as he likes and will never find a beginning nor an end of this surface here. One dimension higher, that's the universe.

Question: And there is nothing outside of the universe?

Landua: I mean, nothing that would be of any relevance for us. I cannot answer the question of what is outside the universe. That is sort of like not really accessible to our measurements, observations, and so on. There might well be something outside, but we have no clue what it is.

Notable Quotes by Rolf Landua

- When I looked at the sky, I saw something which was bigger than my environment, which was sort of a promise, a temptation to explore the space—to find out what I am made of.
- We are failing to mesmerize kids with science, to really make them excited about it. Science is presented as a dead subject, as a bunch of formulas which you have to learn in order to pass the exam.
- All simple experiments have been done many years ago and now we are challenging nature at a scale which is much bigger than it was in the past so we have to really make an effort.
- Basically, we here at CERN have in some sense one big brain which consists of many, many researchers which somehow behave like collective intelligence.

---

11 Albert Einstein (1879–1955)

Albert Einstein was born on March 14, 1879 in Ulm, an important center of industrialization in southern Germany. He was the first child of the Jewish couple Hermann and Pauline Einstein, née Koch; Einstein's father produced electric appliances and lamps. In the following year the family moved to Munich where Einstein attended the Catholic Luitpold Gymnasium until 1894. Some sources say that he was a dreamy and thoughtful scholar. When his father's company went bankrupt, the family moved to Milan. Einstein, who even as a fifteen-year-old was a revolutionary free spirit, chose to leave Munich and his school without taking his exams. The following year, in 1895, Einstein went back to school in Aarau, Switzerland, where he graduated in 1896.

Einstein enrolled at the Eidgenoessische Polytechnische Schule (the "Swiss Federal polytechnic school," later renamed "ETH") in Zurich, with the goal of becoming a teacher in mathematics and physics. He finished his studies with a diploma degree in July 1900. Subsequently, Albert Einstein applied for several university jobs but was unable to secure a position. Finally, in June 1902, he became technical expert, third class, at the patent office in Bern, the capital of Switzerland. In January 1903, Albert married his fellow student, Mileva Maric. At that time, the couple already had an illegitimate child, a daughter named "Lieserl." Einstein's scientific interest was the classic work of theoretical physics, focusing especially on the writings of Ernst Mach, Henri Poincaré, and Hendrik A. Lorentz.

The year 1905 was Einstein's *annus mirabilis*, the miracle year of Albert Einstein. In rapid succession, he published several works on the fundamental problems of physics, which became milestones in the history of science. On June 9, 1905, Einstein's article "On a Heuristic Viewpoint Concerning the Production and Transformation of Light" ("Über einen die Erzeugung und Verwandlung des

Lichtes betreffenden heuristischen Gesichtspunkt") was published in *Annalen der Physik* (*Annals of Physics*). Einstein's article explains the photoelectric effect on the assumption that electromagnetic waves, such as light, may also be described as a stream of small particles. This wave-particle duality is the basis of modern quantum theory. Einstein's theory proved that everything in nature, even electromagnetic waves such as light, is built up from the smallest particles or "quanta." In 1922, Einstein was awarded the Nobel Prize in Physics "for his services to Theoretical Physics and especially for his discovery of the law of the photoelectric effect."

A few weeks later, on June 30, 1905, Einstein submitted another article to the *Annals of Physics*, with the rather unsuspecting title, "On the Electrodynamics of Moving Bodies." In this paper, Einstein delineated the principles of special relativity. The new theory revised the human understanding of space and time by completely rejecting an absolute frame of reference, the so-called "ether"—a continuum previously thought to pervade all space. According to Einstein's theory, space and time were no longer absolute but must always be determined in relation to each other. The dimensions of space and time are mathematically linked to the four-dimensional space-time, or more precisely, Minkowski space or Minkowski space-time (named after Hermann Minkowski, a German mathematician and physicist). As an addendum to his theory of special relativity, Einstein published another article in the *Annals of Physics*, titled "Does the Inertia of a Body Depend upon its Energy Content?"

In this article, the most famous formula in the world, $E = mc^2$, was published. According to this rather short and simple formula, mass and energy are equivalent; they can be converted into each other. They are just two sides of the same coin. According to Einstein's formula, even a very small mass has to contain a huge amount of energy—atomic energy, which was soon to become the focus of the military. Other rules are that with increasing speed, a body becomes heavier and heavier, and the speed of light cannot be exceeded; it is an absolute limit.

Einstein's theory quickly prevailed in the world of science. In 1909, he became Professor of Theoretical Physics at the University of Zurich. During 1911 and 1912, he was with the University of Prague, and in 1913 he became Associate Professor of Theoretical Physics at the ETH Zurich, his alma mater. In 1914, Einstein by invitation of Max Planck, became a member of the Prussian Academy of Sciences and a professor without teaching responsibilities at the University of Berlin. Einstein moved to Berlin, and could, free from any teaching duties, devote his life entirely to science. He began working on his general theory of relativity which he presented at the Prussian Academy of Sciences in November 1915.

The basic idea of Einstein's special relativity theory is that time is relative. It slows down in a moving system. This basic idea was expanded by Einstein in his theory of general relativity; time is not an absolute quantity but a relative one, and time runs differently within systems which move in a relative relationship to one another. In four-dimensional space (three space dimensions plus one time dimension) no straight lines may be present because space and time vary depending on the strength of gravity. From that fact, Einstein deduced that even the light of the stars had to be deflected by the gravitation of the Sun. This effect was confirmed in 1919. Einstein and his relativity theory became world famous. It still determines the physical description of the universe.

Albert Einstein (1879–1955) in Vienna, 1921. (Photo by F. Schmutzer)

Einstein Quotes

- The important thing is not to stop questioning. Curiosity has its own reason for existing. One cannot help but be in awe when he contemplates the mysteries of eternity, of life, of the marvelous structure of reality. It is enough if one tries merely to comprehend a little of this mystery every day. (From the memoirs of William Miller, an editor, quoted in *Life* magazine, May 2, 1955.)
- Not everything that counts can be counted, and not everything that can be counted counts. (Sign in Einstein's office in Princeton.)
- The further the spiritual evolution of mankind advances, the more certain it seems to me that the path to genuine religiosity does not lie through the fear of life, and the fear of death, and blind faith, but through striving after rational knowledge. ("Science and Religion," at the Conference on Science, Philosophy and Religion. © Jewish Theological Seminary, 1941)
- "Only two things are infinite, the universe and human stupidity, and I'm not sure about the former." (From Frederick S. Perls, *Ego, Hunger and Aggression*, 1940.)

- "Science without religion is lame, religion without science is blind." ("Science and Religion" at the Conference on Science, Philosophy and Religion © Jewish Theological Seminary, 1941)
- "I want to know how God created this world. I'm not interested in this or that phenomenon, in the spectrum of this or that element. I want to know His thoughts; the rest are details." (E. Salaman, A Talk with Einstein, *The Listener* 54 (1955): 370–371; *Jammer*, p. 123.)
- "Do not worry about your difficulties in mathematics. I can assure you, mine are still greater." (Letter to a scholar, 1943.)
- "Subtle is the Lord, but malicious He is not." (1921)
- "I have no special talents. I am only passionately curious." (Letter to Carl Seelig, March 11, 1952.)

12 The Japanese Way: Masaki Hori

Masaki Hori is an experimental physicist from Japan. He studied physics at the University of Tokyo, where he received his PhD from the Physics Department in 2000. Since 1995, Dr. Hori has been conducting antimatter research at CERN. In 2003, he won the 19th Inoue prize for young researchers. In 2007, he won the European Young Investigators Award from the European Science Foundation. Dr. Hori is the leader of the antimatter spectroscopy group at the Max-Planck-Institut fuer Quantenoptik (Max Planck Institute for Quantum Optics) in Garching, Germany. At CERN, Dr. Hori works with the ASACUSA experiment (Atomic Spectroscopy and Collisions Using Slow Antiprotons), where he synthesizes exotic matter containing antimatter, like antiprotonic helium. In 2011, ASACUSA succeeded in producing antimatter with a lifetime of more than 1,000 seconds. Using advanced laser technology, ASACUSA is testing the characteristics of anti-atoms, such as charge and mass, to further explore the fundamental symmetry between matter and antimatter. Differences in the symmetry could point to more clues as to why there's so much matter and so little antimatter in the universe.

Masaki Hori (© Michael Krause)

Antimatter I

The universe was created in the so-called Big Bang about 13.77 billion years ago—according to calculations done with the Hubble constant. Due to instantaneous, maybe hyper-luminal expansion, and the associated cooling, a portion of the existent energy converted into matter. In the high-energy collisions of the LHC at CERN, new particles are created, too, together with their corresponding antimatter particles ("pair creation").

According to the experiments, the Big Bang should have produced equal amounts of matter and antimatter—immediately annihilating each other. In fact, the entire universe known to us is made up of matter. However, the question arises: where has all the antimatter gone? Several experiments at CERN (e.g., ASACUSA and ATHENA) serve to explore the matter-antimatter imbalance and its causes.

Masaki Hori's latest prototype of an "antimatter magnet" is made up of two Tesla coils that produce a strong electromagnetic field in which antimatter can be kept "alive" for a longer period of time. Antimatter is produced—and is annihilated very quickly—even on Earth, for example, by violent lightning and cosmic rays entering the Earth's atmosphere.

Antimatter research is also a good blueprint for a movie. Theoretically, it is the ultimate weapon that can destroy everything. In "Angels and Demons," a Hollywood blockbuster released in 2009, the Vatican is going to be destroyed by antimatter that has passed into the hands of criminals. In a German TV movie ("Heroes," 2013), the world has to be rescued after an antimatter experiment at CERN has gone wrong. That is all nothing but pure fantasy. With today's technologies, antimatter can be produced only in very small quantities and only with huge machines, such as those giant accelerators at CERN.

Question: When did you decide to become a physicist? How did your career start?

Hori: I never thought of what I do as a career. The situation right now is that—probably the professional scientists, or the professors—they probably have careers but the young people, including myself, we started it as a hobby. If you think of it as a career, I don't think you could continue for a very long time. Usually, in the system as it is right now, you have to do it as a hobby for, let's say, ten years, and then at the end of these ten years perhaps you are a professional. But the transition of being a hobby to a career is not so clear to me. A lot of people, including myself, start as a kind of hobby, really, and I mean it like this: a lot of students start in the university. Second year, third year, physics seems to be like an awfully difficult subject, and if you talk, for instance, to your parents or to your friends that you are going to do physics as a career [laughs] that really sounds a little bit outside of what is normal.

So I think normally people start it as a kind of hobby and then maybe after a while you start to think of it as a career, but that transition hasn't happened to me yet. I have a job as a physicist but I don't think that I am a career physicist and I try to tell my students not to think of it as a career, not to try to think of it as, "I have to build up my career," because that's not all the time compatible with what I think basics physicists should be. If you think of it as a career it means that you should do what is very fashionable at that time, OK? And the fashionable things are not things which are actually new so there is a trade-off I think each scientist has to look at. I don't know; this word "career" is just a kind of trigger word for me.

Question: A hobby is something you love. Do you love being a physicist?

Hori: You chose a path which is most interesting for you, which is not the career move or something like this. For most physicists it's like this. There is no golden career path that one can take and so it's usually that you really enjoy doing this; and unless you really enjoy doing this it's really hard to continue, especially in the condition that is prevalent now. And this will be equally true for the younger people, students that I teach now. I always say that, "please keep in perspective of why you are doing it," and why is it that you are interested not because it's good for your career or something.

Question: Why did you want to become a physicist?

Hori: That was an important question for me when I was much younger. Now it's like air, and I don't question myself anymore. I don't feel the need to question myself concerning that issue. Now, when students look at me they oftentimes ask me this question, "Why are you doing it?" because maybe they want to find an answer for themselves. I don't think it's something that you can easily articulate to students. You do it, and it is part of your daily life, and sometimes you succeed in the experiment; most often you fail in the experiment, and then you interact with other scientists. Some of them will be hostile to your ideas, some will be very supportive to your ideas, and in this kind of interaction, I think inside there is the answer. But it's not something that you can tell or teach someone.

In my culture, we say that sometimes it's more important peeling potatoes than to think about the universe. And I think in terms of experimental physics, the point is that we try to understand specific things about the nature or the universe. I can only speak for myself, I should say. As an experimental physicist I don't have a grand scheme or a grand philosophy behind my head which makes the whole thing move. I have been very lucky to be able to work with people who are extremely famous in the scientific field, both in Japan and Germany and at CERN, and I have tried to look very carefully at what makes them tick. What makes people make these discoveries which create the Nobel Prize and these things that really make very big progress. My impression was that it is maybe closer to peeling potatoes than to have some over-reaching, extremely large philosophical underpinning. That's my feeling but maybe other people will say other things.

Question: What is the process of peeling potatoes like?

Hori: I think it was Feynman who said that maybe nature is an onion and you peel the layers and there is more inside, and you peel it and peel it and peel it. Inside the nucleus there is the proton, and inside the proton there is the quark and inside the quark there might be smaller levels, and infinitely regressing—and we might in the end get tired of it. You are in trouble if you think that the purpose for doing it is to achieve some ultimate wisdom where you understand everything. People are not able to understand everything. We can only understand a very small part. If someone believes that we understand mostly everything about nature, this is the height of ignorance. There are a lot of things we just don't understand. I think we will never achieve a total understanding within the foreseeable future—within

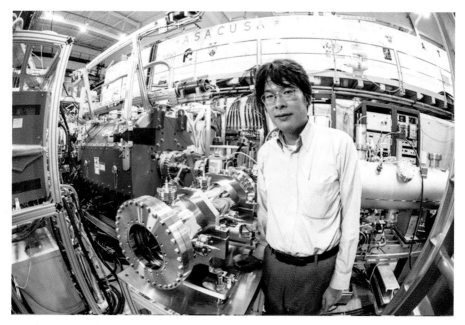

Masaki Hori and the ASACUSA experiment. (© 2012 CERN, CERN-EX-120611610)

my lifetime, within my student's lifetimes, we will not achieve it. The point is that still the whole thing is made out of specific phenomena and you are trying to understand each of them and if you can understand the answer to this one question, then you have progressed the intellectual endeavor of people, of the community, of the human race slightly, maybe. That's your connection I think to the outside. Of course, there is also the issue, "Is what you are doing relevant for society?" And that's a serious question that you must always try to answer. But independently of that, when you are a student, you should not start by thinking of these political issues.

I try to ask my students, thinking of it in a relatively pure way, which is to say you are trying to understand this one bit of nature. If you can understand it a little bit better and you can make other people in the community understand it a little bit better that's what fundamentally the scientific endeavor is about—or this is my understanding of it. A lot of people believe that we are trying to achieve ultimate understanding and I think that's delusion.

Question: Is there a wider horizon out there that we might achieve during the lifetime of mankind or is there another more reachable level that we could reach during our lifetime?

Hori: The fact that the Earth is round and there is gravity, and this huge unimaginable heavy ball is rotating around the Sun—this is almost an ultimate knowledge. Today, students, or children, start already being taught this ultimate knowledge at the kindergarten level; obviously there is an even higher level of knowledge. And in the future, generations of children will be taught about maybe special relativity, about quantum physics; we believe these things are awfully complicated now but I believe in the future even kindergarten children will be taught this. Obviously it seems that there is a progression, but "ultimate" usually means the end, the final knowledge. I don't think we will reach that level and I don't think that that should be the goal.

I think that the natural world is sufficiently complicated that I don't see in the foreseeable future when we will reach this goal. We have to think that there already has been a very large change in the perception of people. For example, I am someone from the other side of the world, down there. Three or four hundred years ago, this was not understandable. This was incomprehensible at that time. You can say that science has not been completely beneficial to us or such kind of things but you cannot deny that it has made a very big difference in the way people perceive things and how they perceive each other. You don't need to be a scientist, you can be a historian, or you can be a normal person, and if you think things in a very fair way, I think you can already see that. It has affected people very strongly.

I don't think the scientists or the people in the last 300 years who studied all these things, that they didn't have it in their minds that they are going to change the world. Not all of them, maybe some of them did. Not all of them had this idea that they will change the world. They just peeled potatoes at their time. I am maybe misrepresenting but maybe a relatively high percentage of them were just interested in some phenomena, like, "Why is the butterfly wing looking like this?" And then they tried to find out and then sometimes with a certain percentage, a certain small fraction of the time, they hit upon something profound and that really had an effect on our lives now. And if you think about all the problems that are occurring now, I think it's necessary that we are trying to find out something new; most of it will be probably useless, but a small fraction of it will be very, very important. And therefore we have to still continue just like our predecessors have done. And I think that's fundamentally the issue. This is my personal feeling about what it is about, what keeps people in this field, moving forward. I think this is the point.

Question: How important is the LHC?

Hori: The Higgs mechanism generates the mass of particles, and most of the theories about this mechanism were developed in the middle of the last century. They are about 30, 40, maybe 50 years old, and until now we had no way to verify the fundamental cornerstone of this theory, which is the detection of the Higgs boson, and this is a pretty fundamental part of the Standard Model. If this part of the Standard Model turns out to be wrong then it's not a minor surgery, it is a relatively major surgery needed to change. The question is—I don't like the word "wrong"—but if we were wrong … in the history of science there are a lot of theories that turned out to be wrong. They were believed for a long time, like, "This is correct, this is correct," but it turned out that nature said, "No, that is not correct!" It is not up to us to decide what nature should do.

Question: Columbus went out sailing to find India and he found America instead. Where does the LHC go?

Hori: It's too early to say. It could be that in the next year or so something new might come up. It is always the unexpected; that is always the fundamental nature of science. Nature is just as it is. It doesn't care.

Question: What do you think will happen in the future?

Hori: We don't know and it is kind of incorrect to say—as a historian I guess—to try to predict what will happen in the future, even in the near future, and decide what is the historical importance of that thing that might happen in the future, which is exactly what people are doing! Suppose the Higgs boson will be discovered.

When Einstein first discovered the special theory of relativity, there was a very famous experiment, called the "Michelson-Morley experiment," where people tried to see this invisible ether in space. They did not observe it but still people believed that theory. Very intelligent, very important scientists believed it.

The Michelson-Morley Experiment

The Michelson-Morley experiment was a physics experiment initially performed by Albert Abraham Michelson at Potsdam, Germany in 1881, and together with the American chemist Edward Morley, with an improved setting, in 1887, at what is now Case Western Reserve University in Cleveland, Ohio. The experiment was set to clarify whether light waves propagate in a medium pervading the whole universe, analogous to sound waves in the air or in water. The Michelson-Morley experiment tested the existence of the "luminiferous aether," or "ether," and its velocity relative to the Earth's movement. The results were negative, and this meant that there was no such thing as the all-pervading ether in space. The speed of light always remained unchanged whether it was measured with or relative to the supposed ether. The Michelson-Morley experiment is considered one of the most important experiments in the history of physics, an *experimentum crucis*. Due to the results of the experiment, the idea of the ether was abandoned. The special theory of relativity by Albert Einstein renounced completely a universal reference system, which included things such as the luminiferous aether.

Question: How is the feeling in the community at CERN about the Higgs boson?

Hori: I came here in 1995 and I always looked at the Higgs boson from the outside because this is not what I did. I am doing antimatter physics. But I saw how there were many rumors that it was to be discovered, and even 15 years ago there were such kind of rumors. In the beginning I was really affected by these rumors. "Oh, it's been discovered." "No, it's not." It is a kind of roller-coaster situation but I think that is very exciting on the other hand. I have a dispassionate part of me which doubts data unless it is really clear, even to outsiders like me. When I am doing antiproton physics, oftentimes I see something and I think this is something new, but it turns out I have been tricked by myself because I am so eager to see something. It helps when a dispassionate person looks at it and says it's OK or not. So from that point of view, I think the Higgs boson is exciting even for an outsider.

I am quite lucky to be here to witness this situation because…Well, the last fundamental particle that was discovered was the top quark. This was in 1995 in the United States; before that it was the W and Z bosons. To be able to be here when a new particle is discovered, that's quite exciting. I have to admit it is exciting even though I am not a person who is doing it.

Question: You made antimatter "live" for ten minutes. How did you achieve that?

Hori: We don't need to achieve that because it is built in. The proton is quite amazing. Why is it that the proton is a stable particle? It's really a complicated object: three quarks are in there, and many, many particles called "gluons." We know that quarks cannot exist at low energies. Two of them together—this is also unstable, they cannot exist for more than ten nanoseconds or so. Three quarks together—then it is called a proton or a neutron. A proton is most stable; we have never seen a proton decay. In the future we might, using very big detectors, but so far all the protons—which are the stuff that make us—are stable. Antimatter has the exact same properties as protons, we believe. Therefore, we don't know why antimatter is stable just like we don't know why the proton is stable. This is why you can maintain antimatter particles if you don't disturb them. If you don't put matter together with antimatter, it will not die. It will exist continuously.

Question: How do you stabilize antimatter?

Hori: Let me say that particles are created from energy. In the very beginning of the universe there was energy only, and the energy transformed into matter but when we try to recreate this in the experiments here at CERN we always find that we also create antimatter in equal amounts. OK? Matter and antimatter are both stable. If you keep them apart, they will remain stable forever, probably [laughs]. Unfortunately, if you have antimatter and you put it into contact with any substance, it can even be the air around us, it will annihilate; it will go back to energy. So the idea is that if you can maintain antimatter in the vacuum without touching anything then it will survive; it will maintain itself.

Question: Is it like a wall of energy that you have to build around it?

Hori: Antimatter behaves in exactly the same way as matter does—except for its charge, but that is a small detail. This means that they behave the same in an electric field, and they will behave in a magnetic field the same. They will be attracted by magnets; in an electric field, antimatter will be repelled from plus, and minus and plus will attract. And by using these effects it is possible to confine charged antimatter in a "magnetic bottle"—that is how it is called. There is a series of magnets designed in a cunning way such that the antimatter inside cannot escape. You can think of matter as being attracted by the magnetic field on the outside, and in this way the particles—without touching anything—will be suspended by the magnetic fields. That new prototype in my workshop works in a slightly different way, which is to say there are radio frequency oscillations which confine the antiparticles. But as I said, this is still under development and we don't know if it works; we might find out in two or three years.

Question: Maybe the model is wrong and there are equal amounts of matter and antimatter?

Hori: We always have to be careful when we discuss about the beginning of the universe. Something happened in the beginning of the universe, and now billions of years have passed. We see some reminiscence of it. We can try to recreate the conditions in an experiment but we don't know exactly what happened.

Now there are scientists who try to recreate the situation by making various experiments but the fact is that we don't know. You have always to understand that we don't know but that we try to recreate the best we can. Now it is true that whenever we try to create matter in an accelerator, we create almost the same amount of antimatter. Matter and antimatter have exactly the same mass as far as we know; we saw that recently with the experiments. It is not so clear to us where all this matter came from but it is not a complicated discussion; it is easy. It is not a philosophical issue; it is a clear experimental fact that when you produce a matter particle you will produce exactly one antimatter particle of the same mass. If you put matter and antimatter together, they sort of cancel each other and annihilate; they will turn back into energy. You can say that in the beginning of the universe there was an unimaginable explosion of energy and that energy turned into matter and antimatter. The easiest way to think about that is, "OK, it must have been created in equal amounts." Don't you agree [laughs]? And if you have matter and antimatter in equal amounts, they will sort of annihilate each other—either you

The detailed, all-sky picture of the infant universe. The image reveals 13.77 billion year old temperature fluctuations (shown as color differences) that correspond to the seeds that grew to become the galaxies. (NASA / WMAP Science Team)

will have nothing or you will have matter and antimatter in equal amounts.

This is why a lot of people tried to find antimatter stars, or antimatter galaxies somewhere. In the universe you have—to some weird effect—antimatter and matter separated. So some part of the universe would be made out of matter, and some part is made out of antimatter. That's plausible, but unfortunately when you use satellites to observe what's called the "gamma background radiation," it is not so. If there were matter and antimatter galaxies somewhere, the two clouds must meet and there would be a lot of annihilations. But if you look into the sky, where is such a source? Nobody can observe such a source.

There is another experiment where you try to see particles of anti-helium or anti-carbon. We all know that matter that makes us was made in the stars. Without stars you can't create the heavy elements that we are made of. That means that if you can see or detect heavy particles of antimatter that means that there must be an antimatter star somewhere there. That's another theory which was proven so far not correct so then we are left with the most, the weirdest conclusion—that the antimatter has disappeared somewhere.

Antimatter II

In his treatise, "Opticks," published in 1704, Isaac Newton wrote that light consists of particles ("corpuscles") that propagate wavelike through the ether. James Clerk

Maxwell realized that light is a type of electromagnetic wave (1864). However, the photoelectric effect—the emission of electrons by substances, especially metals, when light of a specific wavelength falls on their surfaces—could only be explained if light had no wave character but characteristics of a particle (wave-particle duality). In 1905, Albert Einstein postulated in his theory of special relativity that light consists of quanta of light or photons. The photon is a single discrete portion of energy. Light can absorb or emit energy only in integral multiples of these energy quanta. Light (and other electromagnetic phenomena) behaves both as a wave and as a particle (de Broglie, 1924).

In the mid-1920s, Erwin Schrödinger and Werner Heisenberg, two German physicists, adapted the quantum concept, originally developed by Max Planck, to the atomic theory. However, quantum mechanics—as the new concept was called, was not relativistic, i.e. it did not work at very high-energy levels or close to the speed of light.

In 1928, French physicist Paul Dirac came up with a new formula that combined quantum mechanics and special relativity. The new formula characterized the motion of electrons in electric or magnetic fields. In 1933, together with Erwin Schrödinger, Dirac was awarded the Nobel Prize in Physics "for the discovery of new productive forms of atomic theory." Dirac's formula predicted the existence of electrons, which had to have exactly the opposite or symmetrical properties as ordinary electrons have. These particles were later called positrons.

In the 1930s, positrons were detected in cosmic rays by the American physicist Carl David Anderson (1905–1991, 1936 Nobel laureate in Physics). Ernest Lawrence (1901–1958, 1939 Nobel laureate in Physics) built the first true particle accelerator, the cyclotron. His research on mesons and antimatter paved the way for modern high-energy physics.

In 1955, a research group led by American physicist Emilio Segrè (1905–1989, 1959 Nobel laureate in Physics) demonstrated that the antiproton existed—another proof of the fundamental symmetry in nature. Just one year later, the antineutron was discovered. In 1965, a CERN research group led by Antonino Zichichi—and, at the same time, a group at Brookhaven led by Leon Lederman—proved that the nuclei of atoms of antimatter existed (the "antideuteron").

In 1995, the LEAR experiment at CERN detected a small number of antimatter atoms for the first time. The programs for antimatter research at CERN are the ALPHA, ATRAP, and ASACUSA collaborations.

Question: Where did all the antimatter go?

Hori: There are many theories. One of the theories which was pushed by Sakharov. There is an effect called CP violation, a very tiny effect which shows that matter and antimatter are slightly—in certain reactions—not equal. You take the measured CP violation amount that you can detect here at CERN, you plug it into a computer simulation that simulates the universe, and you get an answer which doesn't make sense.

The conclusion is that we don't know where the matter came from or where the antimatter went. This is the status right now. There are various theories that people here are trying to push but none of them agree to what we know or what is seen in experiments. We go more and more into fanciful theories but none of them has proven right. To answer your question: we still don't know and I don't know how we would know in the near future, but you never know. Right now, our models don't agree with how the universe is.

Question: Theory says that the universe once was a singular, very small entity. I can't imagine that. Are there more theories on this?

Hori: There is a famous story—this might be slightly unfair but this is oftentimes said—that Einstein, when he did his theory of general relativity, when he saw his formula, there was a term in there that implied that the universe is expanding—and he erased it because that was not acceptable for him. That was beyond the common sense. And again, we have to be careful that the universe is as it is—irrespective of our common sense. We might think that something huge cannot be contracted into a single point. We might think that all the time and all the space cannot be pushed into a single point.

But that is our common sense and we already know that the laws of nature don't follow common sense always. We are talking here about things which are huge and about time, which is extremely long compared to the common human lifetime or the lifetime of civilizations. So we first have to look at the scientific facts and then we try to reach the most conservative conclusion, which is to say that the universe started as a point and it's contracting or accelerating; the expansion is accelerating. So that's what we believe and that's what agrees with the data best.

You know for the Japanese it's not a problem. Our religion doesn't say it started as a point. Our religion says that our world is going round and round so I don't think you need to compare; you don't need to bias yourself of religious issues or whatever. If it looks similar to your faith then that's your personal issue and I think a lot of people see it in that way. And scientists don't have any problem with that. There is no conflict between the faith of a person and scientific facts. A lot of scientists are religious so we don't see any conflict there.

Einstein's Cosmological Constant

Einstein's cosmological constant, which is denoted by the Greek letter lambda, is the physical constant in his general relativity theory. It denotes the energy density of the vacuum of space. Its value is not fixed; it can be positive, negative, or zero. Einstein added the cosmological constant to his equations so that the universe could be static and homogeneous as it should be according to prevailing scientific models. Einstein's device set his theory in line with the scientific notion of the time; the overall expansion of the universe was completely unknown at the time.

When in 1929 Edwin Hubble proved that the whole universe is expanding—thus it became clear that the universe is not static—Einstein eliminated the cosmological constant from his equations. It is said that Einstein called the cosmological constant "the biggest blunder of my life." Today we know that the universe is even accelerating in its expansion. This phenomenon can only be explained by "dark energy" for now, which acts against an all-pervading gravity that would otherwise eventually crush the universe. Maybe Einstein's cosmological constant, together with our limited understanding of dark energy, are just approximations of the real energy density of the vacuum; it is not empty at all but obviously full of energy.

Question: Is there a common spirit here at CERN? How would you characterize the CERN community?

Hori: This is a uniquely European lab but you would see all kinds of people—thousands of them—working here. I am not from Europe; there are many people from the US, many people from South America, Africa, and so on. The CERN spirit is whatever they bring in and so it's a kind of dynamic situation all the time. I don't think anybody can define what it is and I often find it almost incredible that people of so many nationalities can agree to make one experiment because if you work in a group you will always get into arguments. This is normal.

I think that the nationality or the culture of people is most apparent in the way that they resolve the arguments inside the workplace. I feel that European countries have various different ways to deal with it; the Japanese have one, the

Chinese have one, maybe the Americans have another one, and so oftentimes these manners are very different. I am just shocked how different they are, and sometimes you have to be creative to understand how you can resolve this issue so that we can go forward with our experiments and progress. I did not realize how different people are until I came to CERN. To understand that is part of what you might call the CERN spirit—if there is such a thing.

Question: Could it be a model for society? Could it solve problems?

Hori: It is never solved. There is no real solution to fights, I think. It is just that you need to understand the various cultures in order to understand how they will react. If I say this, then they will react like that. If you are German you have certain expectations; if you are Japanese you have certain expectations and sometimes these expectations are inverted in other countries so it becomes more ... you might call it diplomatic, you might call it psychology ... you are thinking much more when you interact with people. I don't know if that is a natural state of things but that is first of all how it is. I also work in Germany; I work in Munich, in Bavaria, and there is a rule or a set of understanding that people have so they can act much more naturally. If I as an outsider see, for example, some kind of argument in the workplace I have to figure out how is it normal in Germany to solve the problem? I have to work overtime to figure that out but here it is continuously happening for everyone. I think that's the difference. If you work in a foreign country for a very long time that becomes a kind of natural way in that sense that you look at things much more with your brain than your first impulse; because your first impulse is what you learned from your parents or your friends when you were in the kindergarten and this is not what's normal. In Rome, they act differently—that is something that you just have to learn.

Question: Could CERN be something like a role model?

Hori: It is a model. Maybe it is not the best because I think the cultures of each country are coming up from these gut reactions, and if you become too international you are essentially saying that these gut reactions will be suppressed—and maybe you are suppressing part of your heritage as well, so there is always a conflict which way to go. We just have to do it as it is, and I think we have to be perceptive to other people. What are their needs? What are they thinking? This is what people who are working outside of their country always have to think about; if you work in other places of the world, not only in Europe but elsewhere, the customs and cultures will be very different, and so we have to be flexible.

Notable Quotes by Masaki Hori

- I never thought of what I do as a career. The young people, including myself, we started it as a hobby.
- It is sometimes more important peeling potatoes than to think about the universe.
- A lot of people believe that we are trying to achieve ultimate understanding. I think that's delusion.
- Today students, or children, start already being taught this ultimate knowledge at the kindergarten level; obviously, there is a higher level of knowledge.

13 The Nobel Prize Laureate: Carlo Rubbia

"For me, the study of these laws is inseparable from a love of nature in all its manifestations. The beauty of the basic laws of natural science, as revealed in the study of particles and of the cosmos, is allied to the litheness of a merganser diving in a pure Swedish lake, or the grace of a dolphin leaving shining trails at night in the Gulf of California."
— Murray Gell-Mann's speech at the Nobel Banquet in Stockholm, December 10, 1969

Carlo Rubbia.
(© Michael Krause)

Carlo Rubbia was born on March 31, 1934, the son of an electrical engineer and a primary school teacher in Gorizia, Friuli (Italy). After high school, Rubbia studied physics at the prestigious Scuola Normale of Pisa, where he graduated with a thesis on cosmic radiation in 1959. Then Rubbia went to Columbia University (New York) and began experimental work on the decay of muons. Rubbia joined CERN on August 1, 1960 as a fellow. He has been a staff member since August 1, 1961, and at the same time, he has continued his work at Fermilab, Brookhaven in the United States. From 1970 to 1988, Dr. Rubbia was the Higgins Professor at Harvard University.

At CERN, Rubbia continued his experiments on electroweak force with a variety of accelerators, such as the SC, PS, and SPS. Starting in 1981, a completely novel type of accelerator ring, built at his suggestion, started operation: the SppS (Super Proton-Antiproton Synchrotron); it was essentially two alternating gradient synchrotrons with two counter-circulating particle beams. The SppS was one of the first accelerators that made protons and antiprotons collide in the same ring. With this new concept, energies high enough to produce the sought-after vector bosons—transmitter particles of the weak force—were available. The new technology was improved by a research group led by Simon van der Meer, with a technology called "stochastic cooling." Early in 1983, the UA1 collaboration led by Carlo Rubbia eventually discovered the W and Z bosons with the new SppS collider. The 1984 Nobel Prize in Physics was awarded jointly to Carlo Rubbia and Simon van der Meer "for their decisive contributions to the large project, which led to the discovery of the field particles W and Z, communicators of weak interaction."

From 1989 to 1994, Carlo Rubbia was the Director-General of CERN. In 2006–2009, Rubbia was chief scientific advisor at the Spanish Research Centre for Energy, Environment and Technology. Since 2009, Dr. Rubbia has been the energy advisor to the Secretary-General of the UN Economic Commission for Latin America and the Caribbean, and in June 2010, he was appointed scientific director at the Institute for Advanced Sustainability Studies (IASS e.V.) in Potsdam, Germany. Carlo Rubbia has received numerous awards (e.g., the Cavaliere di Gran Croce in 1985, Italy and the Officier de la Légion d'Honneur in 1989, France), 27 honorary doctorates, and he is the author of more than 500 scientific publications. In 2013 Professor Carlo Rubbia was appointed senator for life of the Italian Republic.

Carlo Rubbia is a busy, restless, and always very active scientist who works all over the world to fulfill his commitments as a consultant for large institutions and other state agencies. His secretary is busy canceling his numerous interview appointments because Dr. Rubbia is attending another, even more important conference.

Question: What brought you here to CERN?

Rubbia: I think at the time it was one of the best things you could do.

Question: Has it changed, and if so, in which direction?

Rubbia: That is quite a difficult question. There is a generation gap for sure. I don't know whether things are better or worse, but certainly there is a difference.

Question: Your work today is focused on alternative energies?

Rubbia: That is what I am working on most of these days—new energies. It seems that energy is a physical concept, so physicists should know what energy is, all along. And now with the problems of the climate, the climatic changes and so forth, it is a fundamental problem of society, the burning of more and more fossil fuels. It is a problem; it will destabilize the world. A world without oil will not be a modern world so we must quickly find another solution. So it is a major problem of which I think scientists should be part of the solution.

Question: What would be the direction of this new endeavor?

Rubbia: My direction is essentially research and development. Today, we do not have the means and ways to have an alternative that makes sense, and there is no way—in my view—to live without oil. Now we are reducing it by using more natural gas and coal. I mean, wind is very nice, and photovoltaic is very nice but these are little things. I cannot see "Lufthansa" flying on a photovoltaic aircraft. Therefore we need new matters, new things, and so, once again, like it was in the past, science and technology must come in and invent new ways. And I think there are new ways.

I think the answer will come from research. It seems to me that scientists—like in the past—must do good things for the support of mankind and make an input and intervene and try to see how we can make an input for an acceptable, sensible world for mankind.

Question: Where will the energy come from?

Rubbia: There is a lot of energy, such as energy directly from the Sun or wind energy. There are ways to produce much more energy than we consume today but for using renewable energies, we have to build new channels of distribution to bring the energy from the place where it is generated to the places where it is needed. Therefore we can, for example, use superconductivity to transport energy over long distances without great losses. Whoever is developing technologies in this field today will be ahead of others in the future.

Question: Will there be peaceful use of nuclear energy in the future?

Rubbia: I think so, but in a different form than today. For example, there is nuclear fusion. Practically it is not used yet but we as scientists must always think about the future.

The Rubbiatron

The "Rubbiatron" is based on a concept by Carlo Rubbia. This machine is designed to convert long-lived radio-nuclides, such as plutonium and other highly radioactive materials, into less toxic substances, together with some kind of energy gain. The transmutation or transformation of highly radioactive substances in this nuclear waste incinerator is designed to produce a sustainable energy supply. Rubbia called his conceptual design an "Energy Amplifier," because the energy output of the machine could be a hundred times more than the input.

Rubbia's concept combines a conventional particle accelerator and a fission reactor ("Accelerator Driven Transmutation Technology," or ADTT). In a cyclotron, protons are accelerated to an operating energy of about 1 GeV; the protons will hit a target made up of liquid lead. This process produces neutrons that will hit a mix of thorium and radioactive waste coming from, for example, nuclear reactors. The neutrons will be absorbed by the thorium nucleus. Spontaneously, the thorium will be converted into uranium (U-233) and the uranium will disintegrate; this process releases energy as heat. This heat can then be converted into electrical energy, like in conventional power plants.

The Rubbiatron is a subcritical system, i.e., the chain reaction can never get out of control or become critical. A worst case scenario, like the meltdown of the reactor, is not possible. The aim of the Rubbia concept is the construction of a reactor that provides more energy than needed to generate the proton beam. A first test facility in Belgium, "Guinevere," proved the "ADS-System" (Accelerator Driven System) successful in January 2012. The Belgian beta system is the pilot for the larger system, called "Myra" (Multipurpose Hybrid Research Reactor for High-tech Applications), to be built in 2015.

Question: What does the discovery of the Higgs boson ultimately bring about?

Rubbia: The Higgs boson is the end result of 20 years of very, very hard work. The search for the Higgs is so big and grand, as in medicine, with the search and discovery of a new method to fight cancer. The discovery of the Higgs is a big step in the history of mankind.

Question: What can future generations learn from the search for the Higgs boson?

Rubbia: A lot. You can read about the fact that the Higgs has truly become part of the truth. That they may lead them to pursue further, other researches.

Question: What are the benefits of the search for the Higgs boson?

Rubbia: The search for the Higgs stems from human curiosity. This search has absolutely no practical purpose, first of all. It is equal to zero but we are curious, we want to know where we come from. The curiosity is, in my opinion, one of the most important human qualities.

The desire for knowledge cannot be said to grow in the strict sense, although, as I have shown, it broadens out in a certain way. Whatever grows in the strict sense always remains a single entity. The desire for knowledge does not remain a single entity, it is multiple, and when one desire is satisfied another comes into being, so that, in the broadening out of the desire for knowledge, growth, strictly speaking, does not occur; rather, what happens is that something small is successively replaced by something large.

For instance, if I desire to know the constitutive principles of physical objects, this desire is fulfilled and brought to completion as soon as I know what these principles are. If I then desire to know with respect to each of these principles how it is composed, and what are its sources, this is another and distinct desire. Further, the occurrence of this new desire does not deprive me of the perfection gained through fulfilling the first desire. Such broadening out does not cause imperfection. In the case of the desire for riches, by contrast, what occurs is growth in the strict sense, since the desire always remains one and the same thing; one entity is not replaced by another, since nothing comes to completion and no perfection is attained.

"The Banquet" ("Il Convivio"), IV, xiii, 1–2
by Dante Alighieri (1265–1321), one of the most remarkable
poets of the Middle Ages (*The Divine Comedy*)

Question: What is there to do, besides the search for the Higgs?

Rubbia: We do not know about 95% of the universe! The Standard Model that we have developed over the past decades explains no more than a mere 5% of the universe so there is very much to do for the young generation, and many new things to discover. If the young generation really wants to discover something, chances are huge today. It will certainly not be easy, but the resources are there. The problem is that all the not so difficult things have already been found.

Notable Quotes by Carlo Rubbia

- It is a fundamental problem of society, the burning of more and more fossil fuels. It is a problem; it will destabilize the world.
- It seems to me that scientists—like in the past—must do good things for the support of mankind and make an input and intervene and try to see how we can make an input for an acceptable, sensible world for mankind.
- If the young generation really wants to discover something, chances are huge today. It will certainly not be easy, but the resources are there. The problem is that all the not so difficult things have already been found.

---

14 The American Friend: Sebastian White

> *"Science has explained nothing; the more we know the more fantastic the world becomes and the profounder the surrounding darkness."*
> — Aldous Huxley (1894–1963)

Sebastian White.
(© Michael Krause)

Sebastian Nicholas White studied physics at Harvard College and Columbia University in the city of New York. In 1976, he received his doctorate degree with a thesis on high-energy proton-proton collisions. His academic supervisor was Leon Lederman, who coined the ubiquitous term "God particle" for the Higgs boson (Leon M. Lederman, *The God Particle: If the Universe Is the Answer, What Is the Question?*, 1993). Dr. White is a specialist in quantum chromodynamics. White led the RHIC Zero Degree Calorimeter Project at the Brookhaven National Laboratory (USA) before becoming a project manager for the Zero Degree Calorimeter experiment in the framework of the ATLAS collaboration at CERN in 2006. For part of the year he was at the Center for Studies in Physics and Biology at Rockefeller University. Sebastian White lives in Geneva and New York.

In a *Fox News* article ("A puzzle solved at CERN") Sebastian White comments on his work at CERN:

> Ben Franklin, who spoke good French and was elected (as a physicist) to the Royal Society of London, would have had a great time at CERN and in Geneva. [...] It is possible that there will never again

be a place like CERN where, through a major success of international politics, so many resources and so many people whose primary interest is fundamental research, could have come together.

Question: Where are we?

White: This building is where they built the PS. The Proton Synchrotron is the workhorse for all of CERN. It supplies the protons for the Super Proton Synchrotron, SPS; it supplies them to the LHC and so on and so on, and the magnets were all built in this building. Now it was taken over for the ALICE experiment. What we are doing here is we are developing … the specialty is measuring time in particles, which is a hot topic at CERN these days. We are developing something which will let us see the time structure inside a packet that collides in the LHC. Everything happens within half a nanosecond [10^{-9}], and we will resolve the individual collision and that will be useful for measuring the protons that come out of the collisions very close to the beam.

PS Booster quadrupole prototype, CERN, 1969. (© CERN 1969, CERN-AC-6912088)

Question: How did your family react when you told them that you were going to be a scientist?

White: I am the first member of my family, probably on either side, to be a scientist. I am one of very few who ever went on to get a higher degree. My aunt on my father's side got a PhD at Harvard in English literature, and then she became a nun so on that side there is not much history in science. On the other hand, my mother's family—my mother was born in Holland—her mother's sister, in other words my mother's aunt, married the writer Aldous Huxley [author of the novel *Brave New World*], and that's a family which has been very close to science and continues to be today. And I am quite close to Aldous' grandson, Trevor Huxley; we are friends.

Question: How did you become interested in physics?

White: Radiation, genetic mutation? Something happened. Oh, I don't know. It's hard to explain how that happened. I think the way most people I know—what happens is … we just end up in a library one day because we're curious about something and then you just don't stop.

Question: What were you curious about?

White: I don't know in which order it came in but I was either interested in radios and how they worked so I eventually became an amateur, or I was interested in becoming an amateur so I got interested in how radios work and then how radio waves worked and how light propagates and so on. Something like that, I don't know which one came in first but it was because I was always fascinated by radios when I was 12. So then I read about Tesla, and about Pupin—at that age actually.

Question: What was so interesting about Tesla?

White: He was just one of a number of people like Heinrich Hertz and all those people who were pioneers. I wasn't particularly interested in his personal life, although I read his autobiography. I think I was interested because he was interested in the same things; that was all. And he was a pioneer so I became interested in Tesla, Hertz, and also Faraday—those guys were interesting.

> [Nikola Tesla (1856–1943) invented the alternating current (AC) system that is still used today, and he was a pioneer in radio and X-ray technology. He also invented the Tesla coil, a device that can easily generate very high voltages. Dr. Masaki Hori uses Tesla coils to shield antimatter from contact with matter.]

The Flammarion Engraving

The Flammarion engraving is a wood engraving by an unknown artist. It first appeared in Camille Flammarion's book *L'atmosphère: météorologie populaire*, published in 1888. The image shows a man crawling under the edge of the firmament, who then kneels down and passes through a gap between the sky and the earth. The caption of the engraving translates to, "A medieval missionary tells that he has found the point where heaven and Earth meet."

The Flammarion engraving is considered an exemplary representation of the medieval idea of the Earth as a disc and the overlying heavens. The man, sometimes

Camille Flammarion, *L'atmosphère: météorologie populaire* (Paris, 1888), p. 163.

described as "pilgrim," permeates the celestial sphere and sees the mechanics of the so far unknown world behind it, the pre-images of a celestial mechanism in the universe. The man is presented as the discoverer of the mechanics of the universe, and he is retrospectively considered to be at the beginning of a path that will lead him to current knowledge about the universe and quantum mechanics.

In 1887, Camille Flammarion founded the French Society of Astronomy and became its first president. Flammarion carried out research on parapsychology and was cofounder of the French Theosophical Society.

Question: Tesla was interested in the energy of the universe. Are you interested in that too?

White: Yes, we are all interested in that because there is all this confusion in cosmology these days about dark energy. There are many mysterious things, like vacuum energy and so on, and we don't really understand them, I think. But if you mean: can we sap the energy out of the vacuum and put it into running our cars? I am not that interested. I don't believe it.

Question: When did you decide to become a particle physicist?

White: That happened very late; that's funny. I had an interest in science very early. I was fortunate to be in a good public school system in New York but then a funny thing happened when I was 14, when I was totally into math and radios, somehow, my family got interested in the school that they were starting in New Hampshire. It was a brand new idea to do a school that was sort of based on the teachings of Sir Thomas More [Thomas More School]. Teachers were Catholics, all of them, but none of them were priests or in the clergy anyway. It was very idealistic but was very strong in sciences. In fact it was hard to learn anything in the school except what I read on my own but there were many things that were interesting to me about the school, and maybe I am glad I did it.

And then it was a freaky thing that I went to Harvard from that school. When I went to Harvard I didn't know that I wanted to be a scientist. I thought I might but I didn't know what it meant to be a scientist so I spent a lot more time in philosophy courses and things like that than I did in science courses. Then, just when I was finishing my junior year at Harvard, I went to a teacher and I said, "You know, I am interested in doing some experiment." And he said the way these courses work is there are set experiments. You repeat an experiment, famous experiments that have been done. You get to do three of them during your course. But I said, "I am kind of interested in this experiment that was done recently which involves superconducting tunneling, called the AC Josephson effect. It has a lot of interesting things, it can be used to measure the constants of nature very precisely," and so on.

He said, "If you want to do that, do it; you take as long as you want. You can go around and ask the different graduate students and all the resources that we have at Harvard to do that experiment." A brand new experiment, it had just been done. And that is what I did. I spent a whole semester doing that experiment and it worked. And then he was doing an experiment in an accelerator that Harvard had built; it was still running then so I became like a groupie. I just went to the accelerator at night and hung around the experiment and talked to people and so I was almost finished with college when I decided what I wanted to do physics actually.

Question: You said you didn't know at that time what was required to be a physicist.

White: It'd better say, "What kind of life is it? What do we do?" I knew all about art because my dad was a sculptor. I talked to a friend recently; he is almost finished writing a book about the Harvard experience in those days. And after I talked to him about it, he said, "Your experience is not unusual at Harvard. Harvard is a very difficult place in a way, always has been. It doesn't really take care of you in a sense. My idea of Harvard has always been, if you got in here, you must be pretty good. So now you are on your own, we don't need to help you." And so it's a tough way to do it.

My friend said that surprisingly a huge fraction of the people from Harvard—when he interviewed them later in life, said they didn't really know what to do until they left or were about to leave. I think looking back on it that's what we were all doing in college. We were really wanting to understand what a particular kind of life it would be if you led that life—if you're an artist, if you're a politician, or whatever. So I can't say, "What does it mean to be a scientist?" because I know now because I do it, right? I sort of figured it out because I hung around with the scientists who were doing the experiments at the Cambridge Electron Accelerator [CEA, 1962–1974] but could I put a name on it? Could I describe it?

There is a famous quote from Richard Feynman, you know runners. In America, there are a lot of people who like to jog and run. They get pleasure out of sweating. So Feynman said, "I get a kick out of thinking." And that's pretty much true for scientists. You like to think it out, figure it out, and put it all together. One aspect of it is being irresponsible. Just doing what you want and following what you want; that's true to many scientists. So it's immaturity; you don't do what you're supposed to do, you just follow your thing. Scientists think radical. If they would stop doing so, it would be the end of science.

Question: Is it a kind of instinct?

White: It's a way of working that you learn, of approach and things. I don't think it's instinct. I mean, you have an instinct that draws you to it but it's not instinctual that you naturally have a gift or something. You learn it. But you learn it because it's a match with your personality, or something like that.

Question: Is it a certain way of thinking?

White: I'll give you an example. There is in my community in Long Island the most famous hedge fund manager, the most successful of all time. He's called "Jim Simons" [James Simons, "Renaissance Technologies"], who was a mathematics professor at Harvard and my chairman when I studied math. Then he did a little experiment; he ventured into business. He started a little pocket calculator company and he made some money. And then he decided to make a company and invest his money, and he formed Renaissance Technologies. He says that he would never ever hire a professional Wall Street or money person to work for that company. He'll only hire mathematicians, physicists, astrophysicists, and so on. And then he said, "Why?"

He said that the thing he found valuable when you're trying to make money in the way he was trying to make money is that physicists seem to be a little funny. It's very hard to get them all in a bandwagon to like just say, "Oh this is the hot stock, let's all buy this." They are always questioning. And for him, that was the secret to his company. And he made a lot of money; he made 11 billion dollars in personal wealth or something. So is that a characteristic of physics? I don't know. But that's both a good American characteristic and a scientist's characteristic.

But it's not always true in science. There is another famous story of the German bomb project—I am talking about World War II and Heisenberg, because there was too much reference to the leading scientists, Heisenberg made a misestimate of how much uranium it would take to make a fission bomb. He was wrong, and there wasn't this kind of questioning, this antagonistic, healthy thing. The project failed because people started from a wrong piece of information.

I would say for me, as a scientist, what's good in science is that you question. I believe it's a scientific characteristic but you may say it varies from culture to culture.

Question: Too many scientists won't agree to one line?

White: You can't sell them the Brooklyn Bridge. It's hard to sell them the Brooklyn Bridge.

Question: But here at CERN are they all working towards one big thing?

White: It is a little bit more corporate here at CERN. I mean, it varies from individual to individual but as an organization, CERN does not have this because it is an international organization. It has less of this kind of a chaotic quality that in some cases is very important for science. The Los Alamos project was a kind of collection of very independent and reverent thinkers. CERN is more corporate. It is not the same.

Question: Is CERN kind of unique in the world?

White: It will probably never ever happen again that you put 10,000 people in one place for such an important and such a high cause so it is unique. I mean, it is a little bit inaccurate to say "corporate" because everyone is basically motivated by sort of a higher cause. It is self-motivated in a way, you could say, but of course organizing teams of thousands of people is not trivial. It's hard to do that and not seem bureaucratic.

Question: You are part of this community. What makes you tick in this community? What keeps you going? How does it feel here?

White: Here are 10,000 interesting people; some are more interesting than others. It is a fantastic opportunity because there are many associations you'd be interested in—working with someone who has got special skills, so it is like a big playground. For my entire career, it has been a back and forth between whether it is more interesting to work in big teams or little teams. It's an issue I think in physics. If you only work in little teams I think you're crazy, I'll leave the field. But if you go back and forth between working on small, intense projects with few people, and then going to these bigger sort of corporate projects, international projects then it's OK.

The thing that we found is if you pull together, if you get all the resources that you can get from a few countries and a few universities and put them all together, you can accomplish a lot more. But it is not as fun really as doing something like we are doing this week. Just building stuff in the lab here and running into the test beam at 11 o'clock at night and breaking everything open and getting some results—that's much more fun, especially because you know all the people. You work together not because you are in some enormous collaboration but because you respect each other and have a common interest. I think that's an important thing; for me, it is important to maintain the two sides: the small team work and the big team work. And it is harder and harder these days to have the small team opportunities in my field.

Question: Why is it getting harder these days?

White: Because we know so much. You can't do what Curie or Hertz or Tesla did in the laboratory with much chance of having an impact anymore these days. You have to get pretty lucky because we know too much. The big questions are very hard to answer; we know a great deal already.

Question: What are the big questions?

White: One big question is what we were talking about before, about energy. You see, the Nobel Prize in Physics this year was won for the discovery that if you go very far away, and then looking at the light from the supernova, there are two things you think that might have happened. If you go farther and farther away then you're going back in time; you're looking at a supernova that was sending its light earlier, maybe billions of years ago, or hundreds of years ago or something, so you could get a trend, you could make a trend that this expanding universe from what you think was the Big Bang is either slowing down or actually collapsing, or something like that. And it didn't turn out that way. It appears as though it's not as it's slowing down the way you'd expect when you get very far away. Now, there are different ways of interpreting it. It could mean something very different from what people say. It could mean that we just don't understand gravity when we go that far away. Nobody says—unless we know experimentally—that the force of gravity has to be such and such when you get to huge distances. But a more popular picture is that there is something in the vacuum; it's a vacuum energy called "dark energy" which is pushing, so the force of gravity is there, and the pushing is there. So we mix both concepts that gravity is acting way out there and there is this pushing.

That is an enormous mystery and I don't think we will ever discover it with our experiments here on Earth or let's say in the collider at CERN. So that's a tough one. The one that's probably more in reach is this thing called dark matter. Dark matter is an experimental observation that you can't explain the rotation curves of different stars as you go out in a spiral galaxy without having to introduce some other source of gravitation than the stars you see out there so that's a mystery and it's a large fraction of the gravitational force in our galaxy, for example. That's an interesting one because many people think that it is in reach of the LHC. What is that stuff? It is not in our tables of quarks; it is something new.

The Crab Nebula — Type II Supernovae

The Crab Nebula (Messier 1, NGC 1952) is located in the constellation of Taurus. [The Messier catalogue is an index of 110 astronomical objects, mainly galaxies, star clusters, and nebulae, compiled by French astronomer Charles Messier (1730–1817). The New General Catalogue of Nebulae and Clusters of Stars (NGC) is a catalogue of galactic nebulae, star clusters, and galaxies compiled by John Louis Emil Dreyer in 1888. The catalogue contains 7,840 celestial objects.]

"The Crab" is approximately 6,500 light-years away from Earth. The nebula is five light-years across and is the remnant of a star that ended in a spectacular

The Crab Nebula (NASA, ESA and Allison Loll/Jeff Hester (Arizona State University). Acknowledgement: Davide De Martin (ESA/Hubble)

supernova explosion. This extraordinary celestial event was observed and recorded in the year 1054 AD by Japanese and Chinese astronomers. The Crab Nebula is one of the best known and most studied objects in the sky. It got its name from a drawing published in 1844 by the Irish astronomer Lord Rosse, in which the filaments of the nebula resemble the legs of a crab.

The image was taken with the NASA/ESA Hubble Space Telescope Wide Field and Planetary Camera 2 (WFPC2). It was assembled from 24 individual exposures, and it is the highest resolution image of the Crab Nebula ever.

The fiber-like filaments of today's nebula are the ingredients of the star torn apart by the violent explosion. They consist mainly of hydrogen. In the center of the Crab Nebula, there is a barely noticeable, tiny neutron star. This ultra-compact neutron star is about 10 to 30 kilometers in diameter, according to different data. The star radiates enormous amounts of energy, pulsating in a broad range of radiation. This type of star is called a "pulsar." The Crab Pulsar is one of the most powerful ever observed. It has a spin rate of about 30 times per second, emitting electrons at almost the speed of light, which lets the filaments of the nebula glow. The magnetic field of the pulsar is about 10^8 (100 million) Tesla; for comparison, the LHC achieves a magnetic field of about 8.3 Tesla.

The Crab Nebula is a spectacular demonstration of the devastating effects of a supernova. However, the Crab Nebula is a Type II supernova, where strong hydrogen lines can be detected in the spectrum. In Type I supernovae, no hydrogen lines are present. In contrast to a Type II supernova such as the Crab Nebula, the mechanism of Type I supernovae is thermo-nuclear.

Type Ia Supernovae

Type Ia supernovae are astronomical events that are used today as the most reliable measurements to determine distances in the universe, because they always shine with the same luminosity. They are classified as Type I if they have no hydrogen lines in their spectra. With their help, the nature and appearance of dark energy, and the expansion rate of the universe are examined.

Type Ia supernovae typically occur when a white dwarf star in a binary star system—the burnt-out, very compact remnant of a normal, Sun-like star—accretes more and more matter from his larger companion star until it reaches densities and temperatures high enough to fuse the carbon and oxygen in its core into heavier elements. After reaching a critical mass (approximately 1.4 times solar mass), the white dwarf explodes in a nova.

Type Ia supernovae produce explosions with consistent luminosity because of the uniform mass accretion process. The absolute similarity of this value allows for using Type Ia supernovae to be used as "standard candles" to measure the distance of the cataclysm to planet Earth. The research into Type Ia supernovae is relatively new (1999) and its consequences have been sensational; according to these studies, the universe has been expanding for about six billion years after an eight billion year phase of slowing down—and the acceleration is now increasing.

The two projects: CANDELS (The Cosmic Assembly Near-infrared Deep Extragalactic Legacy Survey) and CLASH (Cluster Lensing And Supernova survey with Hubble), examined Type Ia supernovae to study the properties of space-time and the expansion of the cosmos, and they found that "The Universe has proven to be far more intriguing in its composition than we knew" (CLASH: An Overview,

Supernova 1994D (SN1994D) on the outskirts of galaxy NGC 4526. The image was taken by the Hubble Space Telescope. Supernova 1994D is the bright spot at lower left corner of the image. (© The Hubble Key Project Team/the High-Z Supernova Search Team, HST, NASA)

Astrophysical Journal Supplements, January 4, 2012). This means that what we see is not what is out there. About 23% of the universe is made up of dark matter and about 73% of the universe is made up of something that is driving an accelerated expansion. This "something" has been dubbed "dark energy."

Adam Riess (Johns Hopkins University, Baltimore, USA) is the scientific leader of the CLASH project; Saul Perlmutter heads the Supernova Cosmology Project (SCP) at Lawrence Berkeley National Laboratory and is a professor of physics at the University of California, Berkeley. The teams led by Adam Riess and Brian Schmidt, and the Saul Perlmutter team published their results almost simultaneously. The 2011 Nobel Prize in Physics was awarded to Saul Perlmutter and to Brian P. Schmidt and Adam G. Riess "for the discovery of the accelerating expansion of the universe through observations of distant supernovae." The matching results of the survey were quickly adopted by the worldwide science community and thus determined the direction of further research within astronomy and cosmology, as well as in particle physics.

Question: Dark energy is a big mystery. Are we curious enough to look into this mystery?

White: Well, we do, and then we are also very practical. There are some mysteries we don't really know how to get at them, right? Can we prove telekinesis and all these things? Some things we just don't work on because we don't know how to do it. Dark Energy? I don't know how you deal with that. There are people who have programs … but you could also say, "What does it matter?" I mean, I don't feel that way but you could say that.

OK, we have explained all of physics on earth. You know, all the things that we do care about, which means understanding biology to some extent and so on. What does it matter if there is some other matter that doesn't seem to interact with anything, and it's out there, and it's making gravity? It doesn't matter on Earth. It matters over big distances where we have a lot of it, see? You could just be happy and say, "Well, we have understood enough to make all the progress that we need to make an energy production and solving disease and all that kind of stuff." But then you wonder because maybe it is related to some fundamental thing in our theories, like supersymmetry for example, that's what they say. So it's a mixture of being practical and being curious. You can be curious about a lot of things but what's most interesting is when you figure out how to do an experiment which can get somewhere, and maybe address it. And I think CERN is that kind of place, luckily. I'm afraid that in some let's say "public interest" in science, it's not communicated well.

What is physics? Physics has always been an experimental science. You can have all the theories that you want, and if the theory doesn't agree with something that you measure, it's junk; you throw it away. That's something like an ancient Greek theory about how the Sun goes around the Earth or whatever; it's useless, throw it in the garbage! And there is too much—in the way science is communicated— respect for the theory as opposed to the fact that it's an activity. And that leads to a lot of misunderstanding of science. You get into all these silly debates about evolution and whether it should be taught in America because people say, "Oh, it's just a theory, and there are many theories." It's nonsense. It's an experimental science. We know an enormous amount about the evidence for evolution and to say its equivalent to some hairy brain theory or even some literal interpretation of the Bible just means that you do not understand what science is.

That is what I think is valuable about CERN. It is a place where it's being done, a place where it's being talked about. You know, Aristotelean theory, all this platonic theory—that used to be the way that the world worked when we did not know how to answer the questions. Now we know how to answer them and it's much more exciting. It's more exciting because it may lead to a really good theory that's correct. That's a better theory than just nice to talk about.

Question: What is science for you? What do you get from science?

White: It's partly this thing, this pleasure of figuring things out as important. It has to do with how you organize the things that you learn. It's an important part of it but the other aspect that is important is what I'd like to call the "irresponsibleness." No one tells you that it has to be a certain way or that there's a right way to do things or the right answer. I think that the key to science is independent thinking; that's science to me—independent thinking, which is very poorly understood in the press. Let's take a former candidate for president for the Republican Party, Rick Perry. It's not right to say that if someone is American they can make up any theory they want. The American principle comes from much [...] deeper American thinkers, like Thomas Jefferson. The most important American characteristic is questioning things. Keep questioning things means that you don't necessarily have a faith-based picture of the world; you have a more scientific-based one. More scientific is more American, I think. People will argue about it, but ... that's part of the irreverence, right? Because there is a kind of dignity of the individual in science. The dignity of the individual is your ability to understand things and to question them and put them together. That is science and that is very sort of empowering, I would say.

Question: The world is in not in a good shape. How could science help?

White: I don't know where to start. I don't know what to do with fundamentalists, extremists, and so on, but I do know that many of our problems are technical, and unfortunately to address those problems, you need a culture that respects engineering and [the] technical scientific approach of problems. Otherwise, I think you have the BP oil spill in the Gulf; these things are forever. Fukushima Daichi.

That's why we are in trouble because in the US—more than in China, I think—we have a decrease in the strength of engineering and all these things which address those kinds of problems. If you would list all the technical problems and solve those, the other problems would, maybe by comparison, not seem so great. Energy, aging, diseases—there are all these diseases that we all could just cure because we got a good way to deal with them.

Question: Bill Gates puts a lot of his own money into curing diseases. He also put a billion dollars to energy research. Why do we need all that private money? How could the problems be solved on an international level?

White: I don't think it will ever get completely clear of private money but of course, hopefully, it will become more of an organized effort by the countries. I guess part of what happens is—and I think it must have happened to Gates—you find out how little money it takes to elevate an enormous amount of suffering and then you just say, "This is ridiculous." The countries can't get organized to address this problem in the world. If you have as much money as someone like Gates, you just give it away. It's irresistible; you have to give away your money. But you can't cure all the problems, right?

I think there are people who think about those big problems: energy and pollution, and so on, who like to talk about what's called "behavior modification." I am not a big fan of this idea. They say you have to teach people to be different. You only drink water that you get in glass bottles that you get back to the producer or something like that but I don't think that's the behavior modification. Can you modify the behavior of seven billion people on earth? Well? Maybe we modify the behavior of governments and education. It is very hard to change the behavior of seven billion people but you could change the behavior of industries to do things responsibly; don't create oil spills in the Gulf of Mexico, things like that. That would have an enormous effect and you don't have to do your modifications to so many people. Just do modifications to the behavior of corporations, that they conduct themselves responsibly and just obey the rules. I think that's important.

Question: Which one of those three things: pomegranate, onion, or walnut would be the best representation of the universe?

White: I refuse to answer that question, because how can you talk about something as complex as the universe in terms of these three things. It's impossible. You don't get anything from this. Where is energy [walnut]? There is no energy in this. Even

Sebastian White: scientists think radical. (© Michael Krause)

words fall short in describing the universe, but this really falls short [pomegranate]. You break it open and there are some seeds in it. OK?

Notable Quotes by Sebastian White

- It is possible that there will never again be a place like CERN, where, through a major success of international politics, so many resources and so many people whose primary interest is fundamental research could have come together.
- If you mean: can we sap the energy out of the vacuum and put it into running our cars? I am not that interested. I don't believe it.
- Scientists think radical. If they would stop doing so it would be the end of science.
- Physics has always been an experimental science. You can have all the theories that you want, and if the theory doesn't agree with something that you measure, it's junk; you throw it away.

15 Friendly Competitors: Sebastian White and Albert De Roeck

Albert De Roeck obtained his PhD at the University of Antwerp while working on an experiment at CERN (UA group), studying dynamics in hadron-hadron interactions. Then he spent 10 years at DESY, the German particle physics laboratory in Hamburg, to make precise measurements of the quark and gluon components of the proton. In the late 1990s, Dr. De Roeck returned to CERN, where he was involved in experiments (OPAL) with the Large Electron-Positron Collider. Albert De Roeck led the CMS Analysis Group and was the Deputy Spokesman for the CMS experiment. He is a member of Exotica and the Higgs Group and since 2011 has been a member of the CERN MoEDAL group.

Albert De Roeck (left) and Sebastian White. (© Michael Krause)

Dr. De Roeck is a professor of physics at the University of Antwerp and has worked on more than 400 scientific publications. De Roeck is currently involved in the CLIC study for a future CERN linear accelerator. He is working on scenarios for a significant increase in the collision energy of the LHC.

ATLAS and CMS—A Healthy Competition

De Roeck: From the beginning—and rightly so—the program followed by the CERN community was that there should be two very strong experiments that should also be supplementary at certain points, of course.

White: There must be two very strong experiments, possibly complementary. I think both Albert and I agree that we will attack in these two experiments the same fields in physics, so to say. There may be, will be, very interesting sub-measurements or aspects of the measurements which contrast each other in the two experiments but we are looking for the same particles, definitely.

Question: What is the difference between the ATLAS and CMS experiments?

De Roeck: They sort of differ in the specific technologies and methodology they are using for hunting these new particles but to a large extent they are using very similar techniques and they will be producing—I believe—very similar results but using different types of detectors.

Different detectors have different systematics. By having two experiments it is very important that if we discover something that it is indeed seen in these two experiments, it will give us a lot of confidence because the exact nature of what we are looking at is not predicted for us. We have ideas but somebody will find something and that will need confirmation and I think that is going to be very important, and I think, therefore, it is really worthwhile to have such two major enterprises as we have here.

White: In the very beginning, the people are very excited about their technologies. They think, "Aha, we have a more clever way to look for one particular mode for detecting the Higgs boson," for example. Then the other experiments develop their technologies. That is a period of enormous excitement and lots of technological development. But during this period, you also reach a stage where you realize all the technical difficulties and you work very hard at it. The initial excitement pushes you through that very difficult period. Then in the end you might say both different approaches achieved a certain maturity. Maybe they were a little bit too optimistic in the beginning but in the end we are all shooting for the same targets but we go by different roots.

De Roeck: An important aspect among a community like us physicists is healthy competition. There is a certain kind of vanity with physicists as you have in other disciplines—you'd like to be better than the other so you really work extra hard. You want to be the best, but in other words, it is a friendly competition.

White: In spite of what Albert says, you win and then you slowly get chipped away at what you are winning. When you have that moment, the laboratory, which in this case would be CERN, is very interested in hearing more details, like, "Are you sure? Are you sure this new result is right? What about the other people? Shouldn't we wait a little bit longer?" Maybe that is healthy but I agree with Albert; the initial feeling is excellent. And you have a few drinks with your colleagues from the other experiment and you say, "Well, I hope you guys come along soon. It is very nice where we are now and I hope you come soon." But in the long run it is quite common that the laboratory will encourage you, before you publish, to let the other team have a chance so that there is really confidence in the results. These are fundamental results and it matters really much that they are right. It matters much more that they are right than who got there first.

Question: So, one wants to play—and win also?

De Roeck: I fully agree with what Sebastian says. It is important to be first but one has to be right. The credibility of the field is depending on that and the whole scientific credibility of the experiment. Being right is the prime goal, actually.

White: It is the old story about the dogs which go and search for the truffles. The theorists are the people who own the dogs and the experimentalists are the dogs. So the theorists come up with some beautiful ideas and the experimentalists are the dogs. They go and they fetch the truffles, and they get to hold the truffles—is it dogs or pigs? The pigs get to hold the truffles in their mouth for just a little minute before the theorists then snatch them away.

Question: Where do you prefer to work—in theoretical or experimental physics?

White: I love to be in the middle of it, right in the middle of the search and to see the first glimmering of these new particles that we are looking for. Of course everyone feels the same I think.

Question: Is the same true for you?

De Roeck: From the beginning—I came as a summer student to CERN, now 23 years ago—I was immediately caught by the virus of what is happening here at that time already. They started off a new machine here and they found these famous so-called W and Z particles, and that was a time of great excitement because they knew something new was going to be produced, and they were waiting for that. I wasn't involved in the experiment but being in the lab was already enough to get excited by that. And I think since then I have always been excited about pushing further, trying to find new things, measuring stuff which no man has measured before and seeing how we can learn from that and complete our understanding of matter and in fact the universe where we come from.

16 Rock 'n' Roll, Beer, Billiards, and Music: Jonathan Butterworth

Jonathan Butterworth.
(© Michael Krause)

Jonathan Butterworth grew up in Manchester and did his A levels at Shena Simon College there. He studied in Oxford, getting a BA in physics in 1989 and a PhD in physics in 1992. He then moved to Hamburg, Germany, to work on the ZEUS experiment at the HERA electron proton collider until 2004. Butterworth made the first measurements of hadronic jet photoproduction and was physics chair of the experiment in 2003–2004. He joined University College London (UCL) in 1995, where he was appointed professor of physics in 2005. At CERN, Dr. Butterworth was convener of the ATLAS Monte Carlo group (2007–2009) and of the ATLAS Standard Model group (2010–2012). (To learn more, you can visit http://twitter.com/jonmbutterworth.)

Jon Butterworth writes for his *Guardian* Science Blog, "Life and Physics" [http://www.guardian.co.uk/profile/jon-butterworth].

> *"The United Kingdom has spent more on saving banks in a year than it had on science since Jesus."*
> — Professor Brian Cox in a BBC interview, July 6, 2012

Beauty Is Where You Find It

Question: When did it all start? When did you decide to become a particle physicist?

Butterworth: It is difficult to remember precisely, but I know that when I was eight or nine, I remember that I and a friend of mine decided to write a book containing everything everyone knew about the whole universe. This tells you something about us probably. We didn't get very far with it but it did tell me a lot about how much there is to know and it also showed that what we meant at that stage was essentially stars and planets and galaxies and things. I guess that kind of thinking lead me to physics to some level.

I know I was very keen on math at school and I remember some moment—I think when I was about fifteen, we were drawing gradients; we had to draw them and then measure them. And the math teacher kind of told those sitting in the back of the class, he said, "What we are doing here is learning differentiation, and if you differentiate x square the gradient is always $2x$," and it was just amazing that all these measurements in these gradients were all the approximations to this underlying mathematical rule. And so that was the first time I sensed that math really connects with the world around us.

So that made a big difference to me. It was the first time that I really connected this kind of beauty of an underlying mathematical rule with the real world and the physical universe and to me that's physics. I don't want to do maths except for what it could teach us about the universe around us. Then the math is really important and that's how I do physics.

Question: What does physics mean to you?

Butterworth: To me, physics is looking at the real world around us and seeing the underlying laws which many times turn out to be really beautiful and simple mathematical rules, the way things behave. Why that should be, I have no idea but it seems to be that way and it's beautiful and increases our understanding of the universe we live in, which I think to me is what the motivation is.

Question: What is the beauty of the universe?

Butterworth: Beauty is where you find it, isn't it? I mean it is beautiful just looking out of the window, the Sun reflecting on the sky, the atmosphere in the leaves of the trees. But if you look at that more closely then there is the beauty of how photons really interact with the atoms, and how that is creating those effects. It seems to me that there is beauty on many levels of the universe and I guess the physics we do here is interested in kind of peeling back some of those levels and seeing what's underneath them. And every time we did that, we seem to find more beauty there as well. It doesn't destroy the beauty when you peel back and it will add another layer to the beauty of the universe.

Question: How deep did we already dig into that kind of beauty? What did we find already?

Butterworth: One of the whole motivations for building the Large Hadron Collider is trying to understand the "electroweak symmetry breaking." That is a complex phrase for what is actually a rather beautiful feature of the way we understand the universe. It seems that our current model of physics that we have at the moment—symmetry is a really important part of that. And we know that there is some symmetry between the weak nuclear force and the electromagnetic force which is respected at very high energies but is broken in our everyday world. And so you can say that if you think of symmetries somehow aesthetically, because it is part of the beauty of nature, then the whole enterprise here is to really understand one of these underlying principles, i.e., how is the symmetry broken? Why is it restored at high energies? I think a very general property of our current understanding of fundamental physics is that symmetry plays a huge role and I think that is beautiful in some sense.

Question: Where in the history of physics is the LHC?

Butterworth: It is a difficult question to answer because we know where we have been but we don't know where we are going so we don't know how much more physics there is to do. Sometimes it seems that we are almost to the end of understanding fundamental physics. That we have an almost complete model, and then every now and then something unravels and it seems that you're only at the way in. I guess we had the Standard Model of Physics basically solidified during my early career and we haven't had a revolution in physics for 20 years. We found new things but nothing upturned from the fundamental picture of what's going on.

I think we are looking over a new frontier here. It is more than just going to a higher energy. This is a special energy scale in nature, which is where we can see the weak nuclear force and the electromagnetic force diverge. When you extend to higher energies you always might find something new. Here, we know that this is kind of an event horizon. Some physics look fundamentally different at the LHC than it does at lower energies, and we don't know why that is. I mean, the Higgs boson is the headline that everyone sees, that's what is supposed to be responsible for that—that kind of change in nature. But the important thing is we are peering over the hedge at some new landscape of physics there. We have done that many times in the past. How far we're along, how much more there is to explore, I really don't know.

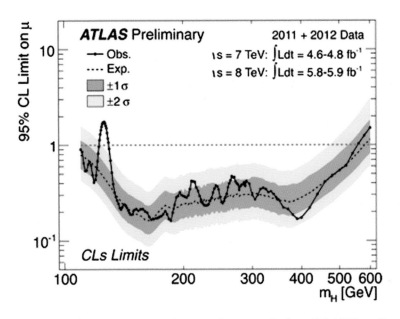

Mass of the Higgs boson, approx. 125 GeV according to results from ATLAS Higgs Search. (© 2012 CERN ICHEP 1459545)

Question: What do you think is behind the hedge?

Butterworth: The Higgs mechanism is a good description of the hedge. What then lies beyond that I really don't know. I am an experimentalist and I would like to see. I really want to see. There are a bunch of candidates there but I suspect they are all wrong. I suspect we actually will find something else or different. But there are all kinds of options. I mean, supersymmetry could show up there. We could find that fundamental particles are actually made out of something else, and the forces that we see are not really fundamental. They might be low energy manifestations of some deeper layer of our understanding. All of that and probably some of it is probably true; some of it is probably not. I don't know. I mean that's why we are looking over the hedge.

Question: What makes you curious? Why are you doing this?

Butterworth: I think testing our understanding of nature is the only way of really being sure that we understand anything, so I think there is lots to understand about studying other people and studying the way society works and it's fascinating to do that. But what really is interesting is the almost inhumanity of the universe.

In fact, that it is not a human construct—that there is some stuff out there. You know, no one ordered quantum mechanics but the physicists. No one said before that the world was as mysterious as that. But yet, if you look at what really happens, you are forced to accept some of those ideas and I really like that clash between preconceptions and the reality that is out there. I think that really changes the way we think and it gives us a more complete picture of where we live, and I can't think of any better way to spend my time.

Question: When you and your colleagues talk, what are the topics?

Butterworth: The desire to understand what's going on around us underlies pretty much everything of what we do. And that translates itself into the desire to actually debug a piece of code as well or understand a bit, our bit, of the detector's working—or to understand a particular calculation. They are a more immediate manifestation of that and you can obviously take real joy in solving problems, even a crossword puzzle can be a lot of fun to solve but I think there is a difference here. I mean problem solving is the feature of many occupations. I think the only difference here is that underneath that problem solving extends all the way to the underlying purpose of why we are here. We are actually not here to make money or something; we are actually here to understand stuff. Day to day, I think the job is very much like any intellectually demanding job that requires solving problems, and they are not all scientific problems. I mean there are political, commercial, personality problems—all that is part of it, but in the end, what we are trying to do is understand stuff and that carries all the way through to the ultimate motivation of the job as well.

Question: What would you say are the characteristics of a proper physicist?

Butterworth: I think there are many, and almost as many as there are physicists. But if I had to say what was the absolute requirement above all of us, it is a certain kind of intellectual honesty. No matter what else is driving you, it could be very good for your career if the result goes one way rather than the other way, but in the end you have to be absolutely honest and pull back from the data a little bit and not kick yourself and not kick your colleagues. And that's very hard, and it's very hard to do when you're all fired up to competition or some particular study but in the end you are betraying yourself and the whole profession. I think that the standard of intellectual honesty is very high and if it ever slips then people stop being physicists in the end.

Question: Is there a common spirit here at CERN? Could it be a role model for the rest of society?

Butterworth: That is something I thought about quite a bit actually. There is a common spirit; it comes from the desire to understand. There are a huge variety of personalities and don't get me wrong, it is not utopian; there are a lot of people who don't get on at all with each other but in the end we have this common ground. We want our experiments to tell us the truth about nature. I think that is the key to why so many different cultures and personalities can work together effectively here. It is because there is this underlying commonality of purpose. It is maybe not a single purpose like "finding the Higgs." I mean it is more like a culture that wants you to understand the universe around us. Now if you can create a similar overriding culture in an organization for the goals of that organization, I don't see why it wouldn't work. I think that what keeps CERN together, despite the many forces that pull people apart, that is the thing that keeps the culture together and that keeps you working.

Question: There are many problems in the world today. How could you, or CERN, answer some of those urgent questions?

Butterworth: I would say it would be misleading to say that the questions we are asking in terms of things like particle physics gear to answer any of these questions; they clearly are not. However, I think that there are many ways the activities here address those questions. The technologies that we developed on the way to building these machines and doing these experiments had many spin-offs. Things like power generation, things like communication. I mean the World Wide Web is the example that everybody brings about. We live in a world in which we are all interdependent. A lot of the innovations at CERN have been about communication and interdependence. The model of the organization that the world can come together for a particular goal is actually quite an important one that should be used to address some of these other problems as well to the extent that it's appropriate.

Question: You used the phrase "cultural conversation." What does that mean?

Butterworth: I think that too often science is seen as a pursuit of a bunch of a kind of priesthood—of people who are different from the rest of society. I think that that's not really true. If I look around in my most cynical mode, if I look around at the newspapers in England, for instance—I am not going to have a go at the tabloids, if you look at the higher end broadsheets, there are pages and pages of book reviews, of discussions of theatre, of discussions of art in general, of philosophy, of course of politics and economics—they impact people's everyday lives. It has always struck me that in the balance there is something wrong.

I think that there is an appetite there for a discussion of science in a more educated form by people who actually understand it. I think scientists at some level dropped the ball on that and decided not to participate in that kind of discussions enough—and when they have, they have been a little bit preachy. And also I think the editors of the newspapers who generally don't have a scientific background underestimate the market for those kinds of discussions. So there is a huge opportunity there in the Internet, in the web that we helped develop, of course. So you can meet people where they want to be met. So it's true you don't want a gooey explanation of quantum field theory in the middle of your breakfast table every morning, but some people do actually. Or some people will see an article that leads them on wanting to know more and if all you have is your paper, you'll never find more. You'll just find the same facile analogies over and over again but if you actually do want to find more, you can go and maybe find it out because the paper has a website and there is a blog there that explains some more—and that points you to the original articles if you want and you can follow the information through to whatever until you had enough.

I think that is a huge change in the way information is made available to people now, and the mainstream media and the big headlines have a huge role to play because they pull people in. I think scientists should be very involved in providing the information that kind of underlies that and the newspapers should be leading people through to that information if they want to. If they don't that's fine—but if they do, they should be able to find it. And I think that is a very interesting area that is developing a lot. I write regularly for the web pages of the guardian in the UK, and there are clearly a lot of people who want to read that kind of thing. It's often not even new stories, it's stuff that is news to the people reading it but for physicists it has been around for a while but it is not widely enough known. [http://www.guardian.co.uk/profile/jon-butterworth]

Question: What do you think people can learn from your blog?

Butterworth: My goal in doing that is not really about physics. It is about getting people to understand science or how science is done or what it means to be a scientist. Because in some places, the perception of science is disconnected and is just another religion or something, and it is fixated in its ideas and some of that criticism is true to some people some of the time. But in general that is a misrepresentation of science. It is really an enterprise of exploring of how the world works and looking for evidence and coming up with ideas, and when they are wrong, working out how you reject them fairly. And I would really like people to see that process. I think leading people through how scientific doubts and lack of knowledge becomes eventually knowledge is a really valuable opportunity. I hope people stay interested in it.

Question: What as the most challenging thing here at CERN?

Butterworth: I have been the physics chairman of an experiment in Hamburg at DESY and I really enjoyed that and I had a lot of friends there. I was well known; I knew everyone on the experiment and I had a lot of fun helping to produce the papers. And when I came here I was completely unknown so that was kind of nice to start again. It was quite challenging, and it was a lot of fun, yeah.

Question: What was the biggest challenge, technically?

Butterworth: When we had no access to the data because of a broken cable somewhere. There were many more challenging technical issues in developing the detector, but, personally speaking, that was the one that was the worst for me. I was maintaining some suffering for a while. Just getting to grips with that and making it work was tough.

Question: If you were a visionary what would the vision be?

Butterworth: I would love us to find something completely unexpected here. I kind of have visions of what I don't want, by the way. In fundamental physics, we think in terms of quantum field theory and people have been playing with string theory, and this and that, and tried to predict anything. I have a feeling that there is something else out there. In a way, thinking of how nature might have run its course—in the same way we went through from classical mechanics to quantum mechanics to quantum field theory. Maybe it's string theory, maybe it's not but I would really like some data to come along that didn't fit into that picture at all and would change the way we look at things. But that's not really a vision. I guess I am not a visionary. I am really an explorer and I don't really have a vision of what I am going to find. I just want to know what's there.

Question: Where are we in the evolution of the Standard Model?

Butterworth: I really don't know. I mean if it would be only the Higgs and only the Standard Model then obviously that would be an evolution and at some level of supersymmetry if it's one of the expected ones. One of the more commonly considered ones would be at some level an evolution of the Standard Model, if it was a new force or something, if we had strong interactions that were responsible for mass generation. That's really a totally different layer of the onion, if you like. There is a whole other set of forces and constituents probably in there that we have no idea about so far. That is why we are doing the experiment.

Question: I understand that nobody understands gravity, right?

Butterworth: That is one of those big clues that we are a long way off from having the theory of everything. I mean that is this idea that at some point we will have the whole of the universe done and dusted and put into equation. Well, I don't really buy into that idea anyway but even if you did, the fact that gravity doesn't fit into our current picture of nature tells you we have a long way to go before we're there. That's one of those things. There are other things—like we know dark matter is there but we don't know what it is and various sorts of things that tell you that there is unfinished business in the understanding of nature. I would hope that at some point the clues of the crossword puzzle will come together and we will fill in some more of the blanks but I am not sure how big the crossword puzzle is.

Question: Is there a place for God?

Butterworth: I think if you want him there is probably a place for him. I don't see a need for him myself but we are never going to disapprove the existence of God here. That is not what we are about. If you choose to interpret our universe through the eyes of a God I am sure you can carry on doing that.

Question: I mean because of the term "God particle."

Butterworth: That's a bit of a nonsense. I mean, you can think of analogies of the Higgs and God. The Higgs is supposed to be the manifestation of some field that fills the universe and our theories don't work without it, and if you are a religious person, you probably think of God. That is one of the attributes of God—that he is everywhere and nothing, makes sense without him. But I don't buy that, and anyway the Higgs clearly does not have anything to do with God.

Question: In 20 years from now, what will the cornerstones of physics be?

Butterworth: We really don't know. This makes people nervous because we have a planning blight on what we want to do in the energy frontier at the moment. It is not at all obvious what the next experiment should be until this one has shown us, and we don't yet know enough to do that so there are proposals for machines: for lepton colliders, linear colliders, muon colliders. It's not clear which of any of those are technically doable, affordable, and address the right questions. We actually have got to know what the right questions are because we don't have the answers from the LHC yet.

There are other areas in particle physics, for instance, in the areas of high-energy cosmic rays and in the area of neutrino physics, where there is an awful lot of interesting stuff going on as well. I was in the big conference in Mumbai a couple of weeks ago where there were really interesting things going on there. There are experiments in Fermilab in America, and in Japan, in the UK where we are starting to look at neutrinos: how they behave and what their masses are, and this is all giving us clues to how the universe came to be and what the laws are there as well, which might complement to what we might learn here at the LHC.

The current round of experiments needs to produce their data before we really know what the next steps should be. And in a way that's science for us. I mean there are very few fields of science where you can map out the progress 20 years ahead. On some level, yes, but if you know exactly what you are going to be doing 20 years from now it's a bit boring, right? I mean we really don't know. We are exploring a frontier and we will try and do whatever seems to be the most appropriate things to address whatever questions are thrown up by the current round of experiments. The LHC itself will continue producing interest and useful data for 20 years but what goes on in parallel, or what takes over after it, will depend on those results.

Question: There is an analogy here, Columbus sailed out to find India and he ended up in the Americas, where are we in our journey with the LHC?

Butterworth: I would say—if you take the Columbus analogy—now we are in sight of the coastline. We are near enough to see whether there are any big mountains near the coast, whether there are any volcanoes or anything. So far, it is pretty much as we expected from the coast but we haven't landed yet and we certainly haven't started to survey to see whether there are new cultures, new civilizations, new minerals, and new resources there. We don't know yet. Maybe it is too harsh to say that we haven't landed. Maybe we are on the beach and we are making a camp now. You know, the sand looks like we have expected; we can't see any volcanoes immediately that hit us in the face but we are sending off search parties in all directions now to try to understand the new territory that we are in.

Question: Does the universe have an end?

Butterworth: Thermodynamics tells us that in the end we will all have no heat, I guess, a long time away. This dark energy idea—the rate of the expansion of the universe is increasing, which is a complete surprise, I am not sure of completely buying it either but it tells you how little we know about the eventual fate of the universe.

Question: Isn't it expanding?

Butterworth: It's certainly expanding; the evidence is there. I doubt I will be there to see, unfortunately.

Question: Is there enough evidence about dark matter that we can say that it is matter—or is it something else?

Butterworth: The question, "Is there evidence?" is always about how long is a piece of string? There is very strong evidence, though, that dark matter is there and that it is some kind of matter, yes. There are multiple bases of evidence from gravity rotation, the cosmological fits, and pictures of colliding galaxies. So there is a lot of evidence that it is matter but that doesn't mean it is a 100% proven so we really understand exactly what it is. There will always be room for some doubts but I would say there is really compelling evidence that there is dark matter there.

Question: Could the Higgs boson be part of it?

Butterworth: No, not really. The Higgs interacting with whatever the matter particle is presumably responsible for giving it mass. But the Higgs itself decays very quickly so it will not be hanging around in the universe being dark matter. If these dark matter particles are weakly interacting massive particles—most people would say they probably are—then it will be that coupling to the Higgs that gives it mass. So it plays a role, but it's not the Higgs directly that is dark matter.

Question: Is there a better picture for the Higgs boson than that particles are traveling through some kind of foam or whatever, and that's how they gain mass?

In 1993, the British science minister, William Waldegrave, offered a prize for the best lay explanation of the Higgs boson. Waldegrave asked UK physicists to explain, "What is the Higgs Boson, and why do we want to find it?" Professor David Miller won the prize. (© 1996 CERN, CERN-MI-9603021)

Butterworth: That is a good enough picture without going through the math. I mean, there is this picture of a cocktail party; it was a colleague of mine who came up with it. That's a good image where someone famous is going through a crowd and gets slowed down by the interactions with the crowd and that's the Higgs. The Higgs field is the crowd so someone you never heard of just goes through because it's massless but someone like the top quark is a famous person and slows down and interacts with the field a lot. And then the Higgs boson itself is the rumor going through the field so there is no person there, it's just a kind of wave in the field itself. There is a clump of people gossiping about something that progresses across the room. It is the best analogy I can think of without bringing in any math.

Question: Is the Higgs boson a field or is it matter?

Butterworth: It's a field. Well, matter is field. In quantum field theory there is not really a distinction. The Higgs is a field; it's a bosonic field.

[Bosons are named after the Indian physicist Satyendra Nath Bose (1894–1974). The British physicist Paul Dirac (1902–1984, 1933 Nobel laureate in Physics) proposed this name for those particles that comply with Bose-Einstein statistics. Bosons are those particles that transmit the forces between the matter particles or fermions. Fermions have half-integer spin; bosons have integer spin. At very low temperatures bosons, unlike fermions, can occupy the same quantum state.]

Question: The Higgs does not have any spin. Why?

Butterworth: We don't know but in order to play the role that it is supposed to do in the Standard Model, it has to be spinless. If you try to spin a spin in there it wouldn't work.

Question: Does it mean that it doesn't move?

Butterworth: The Higgs bosons themselves are moving in the field. The only evidence that the field is really there is how it interacts with particles; it gives them mass. That's the hypothesis so the only direct evidence of the Higgs field is the waves in the field that tells you that the field is there.

Question: Is the Higgs something like glue?

Butterworth: You've got to be careful with such analogies; they can be misleading. You need to write down the quantum field theory because nature at those distant scales isn't common sense. Everything we see around us, we try and think in pictures of what is going on and in the end they can be helpful sometimes. Why do these honey and glue analogies work? It works because the molecules interact with each other. Why do they interact with each other? Because the electrons and the protons interact with each other. Why do they interact with each other? There are probably photons. I mean, you dig that out while in the end you will find the quarks and gluons inside the protons. How do the quarks become mass? Well, we think it is by interacting with the Higgs field. By that stage you already got so far below the level of honey and stuff. Those pictures don't really work at that level but in the end those pictures arise from this level, right? In the end, it is a quantum field describing the quarks, and a quantum field describing the Higgs, and they interact.

Question: You were talking about the "inhumanity of the universe." What did you mean by that?

Butterworth: Finding our place in the universe is a tricky thing. It's not clear we have one. It's not clear the universe cares about us. However, I think somehow by exploring the universe and caring about the universe and how we relate to it, we do give kind of value to ourselves. I think that's a valuable endeavor so I think the universe is inhuman in a sense that we have no particular special place in it but on the other hand we are part of it. So in that sense it is human. We are the human part of the universe.

Question: We discover things and we think the discoveries fit our minds. Do they really?

Butterworth: We try to understand the universe by analogies and we try to draw analogies with things within our own experience. That is a good guide; it is a good way of simulating things in your head but it lets you down sometimes because there is no reason the universe has to behave on very large or very small scales in a way that is comparable to what we see around us in everyday life and in a way that we can kind of intuitively understand. Very often we find ourselves falling back on mathematical descriptions and it is very hard to really picture that beyond the math I think. Some of us do get used to that and you change your ideas about what is normal and you might get some familiarity with the concept which leads at least to the illusion of understanding. To think of things as waves by analogy with water waves or about particles by analogy with a football—it doesn't really work. It is not really what's going on at the quantum level.

Question: In physics there is the saying that the best equations must be very simple to be very good. In reality, is this true?

Butterworth: We have a concept of simplicity and symmetry and beauty and if you like, naturalness—that there are sometimes theories that feel like they have too many bits bolted on the outside. They come not from a simple underlying principle. They have been tweaked and budged, trying to fit the data. And often that's a sign that the original principle was wrong because there was another principle somewhere that won't need all that budging and fixing. So that's the way that we pull progress and think about theory and think about challenging theory with experiment but we always have to remind ourselves nothing is guaranteed. It has been a good guide in the past but that doesn't mean it will continue to be so. We always have to be alert to the possibility that sometimes the universe might just be difficult.

Question: You said that "beauty is where you find it." Where do you find beauty?

Butterworth: I find beauty in a sunny day and a blue sky but I also find beauty in the knowledge and understanding of where the sunlight comes from, and why the sky is blue—and all the myriad underlying things. Understanding that, understanding the principles that lead there as well, simply adds another layer to that beauty. I have no patience with people who think that it is easy to understand how a flower works. It detracts from the beauty; beauty is still often the thing that hits you at the surface and then you see layers and layers and layers of the beauty as well as you understand more and more. And that extends to the kind of physics we are doing at the LHC. Some of the symmetries and the laws are revealed and used every day—they are fantastically beautiful I think.

Question: On your website there is a painting by Vincent van Gogh. Why did you choose that one?

Butterworth: There is a very mundane reason: I like the picture. In the very early days of the web—when I was on shift in Hamburg, it was the early 1990s, one of the first big websites that had any decent pictures was the "Web Louvre." That's how it was called and that was in there so I got it from there and I liked it. But the reason I put it up was that it had a pool table on it and there was a pool tournament in the DESY lab, and I won it. So it was the only picture on the web with a pool table I think. I used to play at school with my friends. We played snooker, actually, on a little table when I was at secondary school. Every Sunday we played; it was ongoing. I think one of my friends was into snooker and I used to go with him and I stayed with it. Pool is a somewhat less serious game and you drink a lot of beer but it's a good way of relaxing.

The Night Café (original French title: *Le café de nuit*), 1888, by Vincent van Gogh (1853–1890).

Question: Are you still with the same group of friends?

Butterworth: I don't see them regularly. They are all in Manchester now, the people I went to school with. Maybe on Facebook, but I don't see them very often.

Question: Where does your fascination with music come from?

Butterworth: I wouldn't say that I was fascinated by music. People ask whether science and music connect. For me, I don't think they do much. I think it is really an antidote to it. Similar to the pool and the beer. I am not one of these scientists who in their free times play chess and read serious books. I tend to read comics and play rock and roll and drink beer and play pool. It is my way of saving my energy for physics [laughs].

Question: Some say now is the golden age for cosmology. True?

Butterworth: There is a coming together of things. Now we understand the laws of fundamental physics to a degree that you can actually use them and you can connect them to fantastic astronomical observations. Things like the cosmic microwave background—a large scale structure of how galaxies are distributed in the universe—and all the information from that are really dates of precision and quantity that cosmologists never had before. There are just huge advances in the field at the moment.

Question: How did research about the cosmic microwave background change our picture of the universe?

Butterworth: The CMB and the way galaxies are distributed in the universe is challenging. Different kinds of dark matter, for instance, would give a very different structure of the universe and then they found recently that the universe is not only expanding but the rate of expansion is increasing. And this is something totally unexpected. It is called "dark energy," which is really a label for something that we really don't understand at the moment but we know it's going on. There is a kind of cosmological Standard Model where all this stuff comes together. It is not as precise as the Standard Model of particle physics; it is not as complete but it is somehow connecting very different observations. Really, there are many challenges and many new insights into the way the universe was formed and how it is developing and what the past of the universe was and what the future might be.

Two galaxies collide. (© NASA/ESA and the Hubble Heritage Team, STScl)

Question: Dark matter and dark energy—you said that you are not quite sure whether you are going to buy it or not. Why?

Butterworth: Science is always a kind of quantified thing. There are very few scientific observations that you believe a 100% but to some point, say 99.9%, you think, "that must be true." Most of what I am used to do within particle physicists is like that. We know that there is a W boson and a Z boson and things. We know what their masses are. The kind of dates we get from astronomy and things is not that precise, and so although there is extremely compelling evidence that there is some dark matter around—you get that from all kinds of different reasons, probably the most compelling one is watching galaxies rotate and collide, so it looks like there is dark matter there. But it's not a certainty at the same level that I am certain that there is a Z boson. The fact that the expansion of the universe is increasing is the sign we have for something that is called "dark energy." First of all, I would say that it is very likely that it's actually expanding and accelerating in its expansion—whether I would call the reason for that dark energy and what that actually means I am really not sure yet. I think it's a label for an open question rather than an answer.

Notable Quotes by Jonathan Butterworth

- Sometimes it seems that we are almost to the end of understanding fundamental physics—that we have an almost complete model, and then every now and then something unravels and it seems that you're only at the way in.
- We are peering over the hedge at some new landscape of physics there. We have done that many times in the past—how far we're along, how much more there is to explore I really don't know.
- We are actually not here to make money or something; we are actually here to understand stuff.
- Finding our place in the universe is a tricky thing. It's not clear we have one.

17 The Higgs Boson — and Then?

On July 4, 2012, scientists working on experiments at CERN announced the discovery of a Higgs-like particle with a probability of 5 sigma. Five Sigma is the gold medal in science. The probability is more than 99.999% or 1 in 3.5 million that the CERN data are correct. With great enthusiasm and dedication, the CERN scientists had evaluated all measurements taken in the years 2011 and 2012. For the historic presentation—it was the first time in decades that a new fundamental particle would be presented—the two teams of the ATLAS and CMS corporations had worked almost around the clock. And they had a reason to hurry: two days before the CERN press conference was held, on June 2, 2012, the office of the US competitor, Tevatron in Chicago, announced that they had confirmed the existence of the Higgs boson long ago but not with 5, only with a 3 sigma. The American institution's attitude might have to do with the fact that the CERN people had set the Higgs presentation on the Fourth of July—a day the Americans celebrate anyway.

The media were enthusiastic about the scientific sensation and praise was great. *The Guardian*, a British newspaper (one of the protagonists of this book, Jonathan Butterworth, writes a highly acclaimed blog for it) compared the dimension of the Higgs presentation with the Americans landing on the Moon, but this time the success was almost all British. *The Economist*, an internationally renowned weekly publication, called the discovery of the Higgs boson "a triumph of humanity." *The New York Times*, an American newspaper, wrote that with this discovery one could finally explain the universe. In France, the media community expressed their pride in the achievement of this magnificent international collaboration; *Liberation*, a liberal daily paper, described the LHC as the modern "Tower of Babel." In Poland, it was expressed that the whole Higgs endeavor was a great triumph of the human spirit ("Gazeta Wyborcza"). Spain ("EL PAÍS") wrote that others should take a lesson from this magnificent collaboration. The German Minister of Science, Schavan, simply called the whole process a "scientific sensation." The Director-General of CERN, Rolf-Dieter Heuer, called the discovery a "historic milestone," and an American blogger ("Borowitz Report") recommended that the CERN people should cheerfully start the production of the new perfume "Hugo Boson."

> *"I did not think that this particle would be discovered within my lifetime."*
>
> — Peter Higgs, July 4, 2012

The press was just as sure as all of the other parties; the discovery of the Higgs is a huge step in the history of the sciences. The Higgs boson is the elusive and long sought-after cornerstone of the Standard Model of particle physics. Now it has been proven to exist. Peter Higgs was present at the historic CERN presentation—as well as his colleagues, who had come to similar solutions almost four decades ago. Dr. Peter Higgs is one of six physicists who in three independent groups postulated the existence of a field which is now called the Higgs field. The other scientists were Tom Kibble, Imperial College London; Carl Hagen of the University of Rochester; Dr. Guralnick, Brown University, Rhode Island; and François Englert and Robert Brout, Université Libre de Bruxelles.

Peter Higgs. (© CERN)

The enthusiastic response to the discovery of the Higgs boson has shown the whole world's interest in the great enthusiasm of the CERN scientists, their experiments, and their discoveries. This book is meant to give the world of scientific research at CERN a face. Hopefully, we have become acquainted with the protagonists of this scientific thriller—titled "To be Higgs or not to be Higgs." Their optimism is encouraging; their open view of themselves as well as their work and tasks at CERN provides insights into how modern science is conducted today. This might be used in finding new work methods, examples, and suggestions in other fields. The international community at CERN could also be a role model for communities, projects, and other complicated tasks around the globe. CERN shows us the way to make our world more understandable and what it means in the name of knowledge to be brave and walk upright to reach a well-defined goal. Might this thinking continue?

All human beings share an interest for the great questions of our short life. Where do we come from? Who are we? Where are we going? Therefore, we realize great trips, great undertakings, and big adventures. In today's physics, possibly on the way to something like New Physics, a new adventure has begun with the discovery of the Higgs boson. Questions are aplenty: is the Higgs alone? Is it part of a larger "family"? What properties does it have? Does it also affect dark matter? Is there really dark energy? Do supersymmetric particles exist at all? Where has all the antimatter gone that must have been there at the Big Bang?

Why is all this of interest to us? Because we are a curious species and we are part of the universe. We are all made of the same elementary particles whose origin lies in the vastness of the universe. We want to know how it all fits together—and therefore the scientists at CERN have been looking for the Higgs boson, and they have found it.

The 2013 Nobel Prize in Physics was awarded both to François Englert and Peter Higgs "for the theoretical discovery of a mechanism that contributes to our understanding of the origin of mass of subatomic particles, and which recently was confirmed through the discovery of the predicted fundamental particle, by the ATLAS and CMS experiments at CERN's Large Hadron Collider."

List of CERN Directors-General

1952–1954	Edoardo Amaldi (Secretary General)
1954–1955	Felix Bloch
1955–1960	Cornelis Bakker
1960–1961	John Adams (Acting Director)
1961–1965	Victor Frederick Weisskopf
1966–1970	Bernard Gregory
1971–1975	Willibald Jentschke (Laboratory I)
1971–1975	John Adams (Laboratory II)
1976–1980	Léon Van Hove
1976–1980	John Adams (Executive DG)
1981–1988	Herwig Schopper
1989–1993	Carlo Rubbia
1994–1998	Christopher Llewellyn Smith
1999–2003	Luciano Maiani
2004–2008	Robert Aymar
2009–present	Rolf-Dieter Heuer

Bibliography

Amaldi, E., *20th Century Physics: Essays and Recollections*, London, 1998

Amaldi, E., *John Adams and His Times*, Rome, 1985

Amaldi, E., "The Scope and Activities of CERN, 1950–1954," Geneva, 1955

Amaldi, E., "Felix Bloch, First Director-General of CERN, 1954–1955," Geneva, 1984

Biagioli, M., *Galileo's Instruments of Credit*, Chicago, 2006

Bohr, N., Open letter to the United Nations, June 9, 1950

Bryant, P.J., "A Brief History and Review of Accelerators," Geneva, 1984

Bryson, B., *Seeing Further: The Story of Science and the Royal Society*, London, 2010

Čapek, M., *The Philosophical Impact of Contemporary Physics*, Princeton, 1961

Capra, F., *The Tao of Physics*, Berkeley, 1975

CERN — LHC the Guide, Geneva, 2010

CERN, Scientific Policy Committee, Progress Reports, Geneva, 1952 ff.

Charpak, G., "Evolution of Some Particle Detectors Based on the Discharge in Gases," Geneva, 1969

De Roeck, A., "Early physics with ATLAS and CMS," Geneva, 2009

Der Spiegel 24/1952, "Strahlen aus dem All," p. 26

European Organization for Nuclear Research (CERN), press release PR 1–10 (1954) (ff.)

Evans, L., "The Large Hadron Collider, Geneva," 2012

Fermi, L., and Bernardini, G., *Galileo and the Scientific Revolution*, New York, 1961

Feynman, R., *Six Easy Pieces: Fundamentals of Physics Explained*, 1998

Feynman, R., *QED: The Strange Theory of Light and Matter*, Princeton, 1985

Finance Committee, CERN, 1952 ff.

Fraser, G., *The Quark Machines*, London, 1997

Ginter, P., and Heuer, R.D., "LHC: Large Hadron Collider," Baden, 2011

Giudice, G., "Fifty Years of Research at CERN, from Past to Future," Geneva, 2006

Gooding, D., and Pinch, T., *The Uses of Experiment: Studies in the Natural Sciences*, Cambridge, 1989

Hermann, A., Belloni, L., Krige, J., Mersits, U., and Pestre, D., "History of CERN," Amsterdam, 1987

Heuer, R.D., "The Future of the Large Hadron Collider," Geneva, 2012

Holton, G., *Thematic Origins of Scientific Thought*, Cambridge, 1974

Hutten, E.H., *The Origins of Science*, Westport, 1962

Jungk, R., "Die große Maschine," Muenchen, 1985

Kosso, P., *An Introduction to the Philosophy of Physics*, New York, 1998

Lederman, L., *The God Particle*, Boston, 1993

Mach, E., *Die Mechanik in ihrer Entwicklung*, Leipzig, 1883

Papaefstathiou, A., "Phenomenological Aspects of New Physics at High Energy Hadron Colliders," doctoral thesis, Cambridge, 2011

Penrose, R., *The Road to Reality*, New York, 2004

Petitjean, P., "Pierre Auger and the Founding of CERN," Paris

Rammer, H., "Two New Caverns for LHC Experiments ATLAS and CMS," Geneva, 1998

Randall, L., *Knocking on Heaven's Door*, New York, 2011

Randall, L., *Warped Passages*, New York, 2005

Regenstreif, E., "The CERN Proton Synchrotron," Geneva, 1960

Rose, F. de, "Meetings that Changed the World: Paris 1951: The Birth of CERN," *Nature*, Vol. 455, 11 September 2008

Salaam, A., *Gauge Unification of Fundamental Forces*, London, 1980

Sammet, J., "LHC—Beschleuniger," 2008

Schopper, H., "Ein Licht der Hoffnung," *Physik Journal* No. 3, 2004

Scientific Policy Committee, CERN, 1952 ff.

Smith, C.L., *The Large Hadron Collider*, Oxford, 2012

Taylor, L., "Functions and Requirements of the CMS Centre at CERN," Geneva, 2007

UNESCO, Records of the General Conference, Sixth Session, Paris, 1951

UNESCO and Its Program XI, European Cooperation in Nuclear Research, 1954

Watson, W.H., *Understanding Physics Today*, Cambridge, 1963

Weinberg, S., *The First Three Minutes*, New York, 1977

Weisskopf, W., *Knowledge and Wonder*, Cambridge, 1979

Westphalen, T., *Proton-Synchrotrons & Colliders*, 2004

White, S., "Heavy Ion Physics with the ATLAS Detector," Breckenridge, 2005

Zajonc, A., *The New Physics and Cosmology*, Oxford, 2004

Glossary

acceleration: The process of giving energy to charged particles in a beam, achieved by RF (radio-frequency) cavities.

accelerator: A particle accelerator (such as the Large Hadron Collider) is a machine used to produce particles of very high energy with speeds close to the speed of light.

accelerator magnets: Different types of magnets in accelerator (storage) rings. The main accelerator magnets are dipole magnets for bending, quadrupole magnets for focusing, and sextupole magnets for controlling a beam's chromaticity (http://cas. web.cern.ch/cas/Loutraki-Proc/PDF-files/D-Russenschuck/CAS-Russenschuck. pdf).

Amaldi, Edoardo (1908–1989): Born on September 5, 1908 in Carpaneto (Piacenza, Italy). He became one of the "Via Panisperna boys" (Italian: *I ragazzi di via Panisperna*)—the nickname comes from the address of the Physics Institute at the University of Rome. Other members of the group were Enrico Fermi, Oscar D'Agostino, Ettore Majorana, Bruno Pontecorvo, Franco Rasetti, and Emilio Segrè; all of them were young aspiring particle physicists, except for D'Agostino, who was a chemist. Amaldi got his PhD in Physics in 1929. He was the main collaborator with his teacher Enrico Fermi (1938 Nobel laureate in Physics) in his research on radioactivity and the properties of the neutron. After WWII, Amaldi held the chair of "General Physics" at the University of Rome and became one of the founding fathers of CERN. In 1952, Amaldi was elected the first Secretary General of CERN.

antimatter: Material composed of antiparticles. Antimatter particles have the same mass as particles of ordinary matter but opposite charge. Theoretically, there is no difference between matter and antimatter except their charge. When they meet, they annihilate instantly. In a world made of matter, antimatter is very difficult to produce. In 1995, CERN announced the first successful production of anti-hydrogen atoms with the Low Energy Antiproton Ring. Why the universe is composed almost entirely of ordinary matter is an unsolved mystery ("CP violation").

antiproton: Antimatter equivalent to the proton with the same mass but negative electrical charge.

atom: The Greek philosopher Democritus called the smallest unit of matter "atomos" ("not able to be cut"). The center of the atom consists of a dense nucleus surrounded by negatively charged electrons. Almost all of the atom's mass is concentrated in the nucleus, about 99.94%. The nucleus consists of positively charged protons and neutral neutrons. Protons and neutrons are made up of quarks and gluons. According to the simplest model, the negatively charged electrons move in elliptical orbits around the nucleus. According to quantum theory, the exact whereabouts of an electron cannot be determined; the electron's position can only be expressed in terms of probabilities. Therefore, the radius of an atom can also not be determined exactly but it is approximately 10^{-10} meters in size. The nucleus is about 10,000 times smaller. The size of a proton and a neutron are similar, in the order of 10^{-15} meters. Electrons are incredibly small, almost point-like particles. Scattering results very recently imply that their size is around 10^{-18} meters.

ATLAS: The largest experiment at the Large Hadron Collider at CERN. Atlas is an abbreviation for "A Toroidal LHC ApparatuS"; it is a general-purpose detector with which to enter new territory in the exploration of matter (http://atlas.web.cern.ch/Atlas/Collaboration/).

baryons: Massive particles that are made up of three quarks. They are held together by gluons, the carriers of the strong force. Protons and neutrons, as well as other particles, are baryons. Baryons, as well as mesons, are hadrons—particles that interact through strong force. Mesons are distinct from baryons in that mesons are composed of only two quarks. Baryons are fermions, while the mesons are bosons.

beam: (*see* particle beam)

BEBC: The Big European Bubble Chamber began operation at CERN in 1973. The BEBC is a stainless steel vessel filled with 35 cubic meters of liquid hydrogen, deuterium, or a hydrogen/neon mix. The BEBC was operated in the beam-line of the SPS until 1984. During its lifetime, the BEBC had delivered 6.3 million photographs to 22 experiments in hadron or neutrino physics. At the time, the BEBC was one of the largest and most prolific particle physics experiments in the world. Today, the BEBC is on display at the Microcosm Museum, CERN Meyrin site.

Big Bang: According to the Standard Model of physics, the universe began with a cataclysmic explosion around 13.75 billion years ago. The Big Bang theory is the prevailing cosmological model about the origin and early development of the universe. According to the model, the universe began with an extremely small, extremely hot, and extremely dense state. After the initial explosion it began expanding rapidly, and it continues to expand today. "The Big Bang Theory" is the most successful sitcom worldwide. The show is centered on four geeky physicists sharing an apartment in Pasadena, California.

Bohr, Niels Henrik David (1885–1962): A Danish theoretical physicist who developed the Bohr model of the atom and made fundamental contributions to quantum theory. Bohr won the 1922 Nobel Prize in Physics "for his services in the investigation of the structure of atoms and of the radiation emanating from them."

bosons: One of two classes of elementary particles (the other being fermions). The name was coined by British physicist Paul Dirac after the Indian physicist Satyendranath Bose (1894–1974). Bosons are fundamental particles with integer spin, such as the four force-carrying gauge bosons, and the Higgs boson, or composite particles such as mesons and nuclei with an even mass number, such as deuterium.

bubble chamber: A detector filled with a liquid (hydrogen, deuterium, Freon) close to its boiling point, where the ionizing particles' trajectories materialize in the form of tracks made of bubbles. Bubble chambers were the main experimental tools in high-energy physics in the 1950s and 1960s. At CERN, there was the BEBC, and Gargamelle. Bubble chambers are no longer in use because they have been superseded by faster electronic detectors.

bremsstrahlung: In German, "braking radiation." All charged particles that are accelerated emit electromagnetic radiation: The greater the acceleration, the greater the radiation. Electrons will radiate and lose energy much more quickly than protons.

cloud chamber: A sealed environment used for detecting particles. A cloud chamber contains a super cooled, supersaturated vapor. When a charged particle interacts with the vapor, it ionizes the vapor and a mist will be produced and seen as a trail.

CMS: The "Compact Muon Solenoid" is a general-purpose detector to investigate a wide range of physical phenomena (Higgs boson and SUSY). In competition with the ATLAS experiment but working with a different methodology (see interviews with Virdee/De Roeck-White).

collider: A type of particle accelerator, either a ring or a linear accelerator, used to collide a beam of particles with a fixed target or two counter-circulating beams of particles head-on.

collisions: Occur between a beam particle and a fixed target, such as a proton in a bubble chamber or another particle coming in the opposite direction in a collider.

cosmic rays: High-energy particles composed of protons and single atomic nuclei of unknown origin. New data suggest that cosmic rays may originate from supernovae explosions. When they penetrate the Earth's atmosphere, they may produce showers of secondary particles. The term was coined by Robert Millikan in the 1920s. Cosmic rays are the most energetic particles known to exist in the universe, with energies of up to 10^{20} eVs.

cosmology: The scientific study of the origin, evolution, current state, and structure of the universe. The prevailing theory about the origin and evolution of the universe is the Big Bang theory.

cyclotron: A type of accelerator in which particles are accelerated by a radio frequency electric field along a spiral path. Cyclotrons were first operated as early as 1932.

Democritus of Abdera (460–c.380 B.C.): A Greek philosopher. Like his teacher, Leucippus, Democritus postulated that nature is composed of small, indivisible units: the atoms (from Greek "atomoi"; "atomos" = indivisible). Democritus of Abdera proposed atoms as the key to a simple universe: "Only the atoms and the void are real."

deceleration: The process of taking energy from a particle beam ("slowing it down").

detector: Also known as a particle detector. It is a device used in experimental or nuclear physics to identify high-energy particles or their debris in a particle accelerator.

Dirac, Paul Adrien Maurice: A British physicist (1902–1984). His version of quantum mechanics was consistent with Einstein's theory of special relativity. Dirac predicted the existence of the positron (positively charged electron). Dirac shared the 1933 Nobel Prize in Physics with Erwin Schrödinger "for the discovery of new productive forms of atomic theory."

dark energy: A hypothetical form of energy that permeates the entire universe. Based on the Standard Model of cosmology, the universe contains about 68% dark energy. It is the cause for the accelerated expansion of the universe—if the theory is right.

dark matter: Makes up about 27% of the universe. Science does not know what it is made of at all. Dark matter comes not in the form of stars and planets that we see; the most common view is that dark matter is made up of exotic particles like axions or WIMPS (Weakly Interacting Massive Particles).

Einstein, Albert (1879–1955): A German-American physicist who developed the special and general theories of relativity that revolutionized the world of physics in the years after his *annus mirabilis* in 1905. Einstein is considered the most influential physicist of the 20th century. Einstein received the 1921 Nobel Prize in Physics "for his services to Theoretical Physics, and especially for his discovery of the law of the photoelectric effect."

electrical charge: A basic property of elementary particles. There are two types of electric charges: positive and negative. Particles with electric charge interact with each other through the electromagnetic force, creating electric fields.

electromagnetic force (interaction): One of the four fundamental forces, or interactions, of nature. Electromagnetism causes the interaction between electrically charged particles; it is responsible for practically all phenomena in daily life. It is the force that holds electrons and protons together inside atoms, and it is responsible for the intermolecular forces between molecules. It is responsible for the atomic structure, chemical reactions, the attractive and repulsive forces associated with electrical charge and magnetism, and all other electromagnetic phenomena. It is carried by the photon.

electron: The lightest stable elementary particle with a negative electrical charge, a type of lepton and a fermion. It was discovered by Joseph John Thomson (1856–1940, 1906 Nobel laureate in Physics) in 1897. Negatively charged electrons are bound to the positively charged nuclei of atoms by the attraction between opposite electric charges. In a neutral atom, the number of electrons is identical to the number of positive charges in the nucleus. The rest mass of the electron is approximately 1/1836 that of the proton.

electron volt: A standard unit of measurement (symbol eV) commonly used in nuclear physics. It is the energy gained (or lost) by an electron moved across an electric potential difference of one volt. The Large Hadron Collider accelerates protons to an energy of up to 7 TeV (teraelectron volts = 1.000.000.000.000 (trillion) electron volts).

electroweak force (interaction): The unified description of two of the four fundamental forces of nature: electromagnetism and weak interaction. At low energies, these two forces appear to be very different. Above a certain energy level (around 100 GeV) they merge to form the electroweak force. Abdus Salam, Sheldon Glashow, and Steven Weinberg were awarded the 1979 Nobel Prize in Physics "for their contributions to the theory of the unified weak and electromagnetic interaction between elementary particles, including, inter alia, the prediction of the weak neutral current." Two CERN scientists, Carlo Rubbia and Simon van der Meer, shared the Nobel Prize in Physics for the discovery of the W boson and Z boson, the carriers of the electroweak force.

elementary particles: An elementary particle is not known to be composed of other particles. They are the smallest constituents of matter: six quarks, six leptons, the gauge or vector bosons, and the Higgs boson.

event: An event in particle physics is a collision of two subatomic particles in the collision zone of a particle accelerator. The particles are scattered during this event, and a lot of new—or already known—particles may be produced.

Fermilab: "Fermi National Accelerator Laboratory" is one of the most prolific particle research centers of all time and a competitor of CERN. It is "the laboratory committed to firm principles of scientific excellence, esthetic beauty, stewardship of the land, fiscal responsibility and equality of opportunity." Fermilab was commissioned in 1967; the bottom quark (1977), as well as the top quark (1995) were discovered at Fermilab, named in honor of 1938 Nobel laureate Enrico Fermi (http://www.fnal.gov/).

fermion: Named after Enrico Fermi (1901–1954, 1938 Nobel laureate in Physics). A fermion is any particle that has an odd half-integer (1/2, 3/2, and so forth) spin.

Most composite particles, like protons and neutrons, as well as quarks and leptons, are fermions. Fermions are the particles matter as we know it is made of.

focusing: In particle physics, strong focusing (or alternating gradient focusing) is the effect on a particle beam to make it converge.

fundamental particles: Also known as elementary particles, they have no substructure. These are fermions (quark, leptons, and their antiparticles) or particles that make up matter and the fundamental bosons, the particles that mediate interactions between fermions.

gluon: Elementary particles that act as gauge bosons for the strong force (interaction).

gravity: Or gravitation, is the universal force of attraction acting between all matter. It is the weakest of all forces yet it is the dominant force in the universe. It is also the most unknown. The law of gravity was discovered by Isaac Newton. All physical bodies attract each other through gravity. In modern times, the phenomenon of gravitation is most accurately described by Einstein's general theory of relativity.

graviton: The (hypothetical) massless exchange particle that mediates the force of gravitation. It has not been discovered yet.

GUT (Grand Unified Theory): In contrast to the Standard Model of physics, the GUT model merges the three gauge interactions (electromagnetic, weak, and strong) into one single force or interaction. GUT was proposed by Howard Georgi and Sheldon Glashow in 1974. Due to the complexity of any GUT (introduction of new fields and dimensions) there is no generally accepted GUT today.

hadron: In elementary physics, a hadron (Greek: *hadros*, "strong, thick") is a composite, strongly interacting particle made up of three quarks and held together by gluons. Mesons and baryons are hadrons; mesons are composed of a quark-antiquark pair; they have integer spin and are bosons. Baryons (protons and neutrons) have half-integer spin, are made of three valence quarks, and are fermions.

Higgs, Peter Ware: A British theoretical physicist (born May 29, 1929), emeritus professor at the University of Edinburgh. The 2013 Nobel Prize in Physics was awarded jointly to François Englert and Peter W. Higgs "for the theoretical discovery of a mechanism that contributes to our understanding of the origin of mass of subatomic particles." In 1964, Higgs proposed a broken symmetry mechanism in electroweak theory, explaining the origin of mass of elementary particles.

Higgs boson: An elementary particle theorized by Peter Higgs in 1964 and confirmed by CERN experiments ATLAS and CMS in 2012. The Higgs boson explains why particles have mass.

Higgs mechanism: Mechanism explaining how particles gain mass. According to theory, they do so by interacting with the Higgs field that permeates all space.

ionization: The removal of electrons from atoms to produce ions.

LEAR: The "Low Energy Antiproton Ring" at CERN. It was in operation from 1982 to 1996; the first nine atoms of anti-hydrogen were observed in the LEAR experiment in 1995.

LEP: "Large Electron Positron Collider." A circular proton-antiproton accelerator: circumference 26.658 meters and radius 4.242 meters. One of the largest accelerators ever constructed and the most powerful accelerator of leptons ever built. In operation during 1989–2000, with four experiments: ALEPH, DELPHI, L3, and OPAL. In the year 2000, LEP was dismantled to make room for the Large Hadron Collider.

LHC: The "Large Hadron Collider." Largest particle accelerator in the world. "Flagship" of CERN, where 10,000 tons of liquid nitrogen and 120 tons of liquid helium are used to cool down the tubes to –271.3 degrees Celsius or 1.9 degrees Kelvin. Such a cold temperature is required to operate the superconducting magnets. An operating current of 11,850 amperes flows in the dipoles to create a magnetic field of 8.3 Tesla—100,000 times stronger than Earth's magnetic field. Collision energy (center of mass-energy) up to 14 TeV. The internal pressure of the LHC is 10^{-13} atmospheres, ten times less than the pressure on the Moon. Started in 2008, it will live until 2030 (https://www.google.com/maps/views/view/streetview/cern/cern-large-hadron-collider-tunnel/v8rqarsznUUAAAQJODm7NA?gl=de&heading=357&pitch=78&fovy=75).

LHCb: The main purpose of the LHCb detector ("b" stands for a heavy quark known as "bottom") is to investigate nature's preference for matter over antimatter ("CP violation"). (*See* interview with Tara Shears.)

lepton: There are six leptons altogether. These subatomic elementary particles can have electrical charge (electron, muon, and tau) or not (neutrinos). Neutrinos have very little mass and are very hard to find. Leptons respond to the electromagnetic force, weak force, and gravitational force; they are not affected by the strong force.

luminosity: In accelerator physics, luminosity is the number of particles per unit area, per unit time multiplied by the opacity of the target. Luminosity is an important value to characterize the performance of an accelerator. The LHC features a design luminosity (collision rate) of up to 10^{34}, per square centimeter, per second.

mass: Or inertial mass is a property of a physical body, or particle, characterizing the body's resistance to being accelerated. According to the theory of special relativity, mass and energy are equivalent ($E = mc^2$).

meson: A meson is a hadronic subatomic particle made up of a quark and an antiquark bound together by strong interaction. Mesons are unstable, with a lifetime of only a few hundred milliseconds. They appear in nature only as short-lived products of high-energy interactions, for example, in cosmic ray interactions in the Earth's atmosphere.

neutrino: A fundamental, electrically neutral particle (in Italian: "small neutral one"). The neutrino has half-integer spin and is therefore a fermion. Three types of neutrinos are known: electron neutrino, muon neutrino, and tau neutrino. They have a very tiny mass and are affected only by the weak interaction. They are therefore able to pass through great distances in matter without being affected by it. The solar neutrino flux on Earth is about 65 billion neutrinos, per square centimeter, per second. The existence of neutrinos was predicted by theoretical physicist Wolfgang Pauli in 1930; in 1956, they were detected experimentally.

neutron: A subatomic particle without electrical charge. Along with protons, neutrons make up the nucleus and are held together by the strong force. The neutron is composed of three quarks: one up quark and two down quarks. It is therefore a baryon. In 1932, James Chadwick proved the existence of the neutron. He won the 1935 Nobel Prize in Physics for his discovery.

nucleon: Collective name for the two particles that make up the atomic nucleus: the neutron and the proton. Both are hadrons, and baryons, made up of three quarks.

particle beam: A stream of charged or neutral particles. A beam of charged particles consists of electrons, positrons, protons, or atoms, such as lead. A beam of charged beams may be further accelerated by use of high resonant cavities. Today, the charged particles are accelerated to almost the speed of light. Unbunched beams have no substructure whereas in bunched beams, particles are distributed into bunches. In the context of bubble chambers or other "fixed target" experiments, protons coming from an accelerator are used to produce "secondary" beams (containing pions, kaons, or antiprotons) whose interactions are then studied in bubble chambers.

Pauli exclusion principle: Named after Wolfgang Pauli (1900–1958), who was awarded the 1945 Nobel Prize in Physics for his "decisive contribution through his discovery of a new law of nature, the exclusion principle or Pauli principle." The Pauli exclusion principle means that no two identical fermions (electrons, protons, or neutrons) may occupy the same quantum state simultaneously. Bosons (photons, mesons, the four gauge bosons of the Standard Model, and stable nuclei of even mass number, such as deuterium) are not subject to the principle.

photoelectric effect: The emission of electrons by substances, especially metals, when light falls on their surfaces. The photoelectric effect suggested a particle nature for light. The effect was discovered by German physicist Heinrich Hertz in 1887. Further study of the effect led to significant steps in quantum physics. In 1905, Albert Einstein described how the photoelectric effect was caused by absorption of quanta of light. This earned Einstein the 1921 Nobel laureate in Physics "for his services to Theoretical Physics, and especially for his discovery of the law of the photoelectric effect." In 1926, American chemist Gilbert Newton Lewis coined the term "photon" for the quantum of light and all other forms of electromagnetic radiation.

photon: An elementary particle with zero rest mass. It is the force carrier for electromagnetic interaction. Photons exhibit wave–particle duality, having properties of both waves and particles.

Planck, Max Karl Ernst Ludwig (1858–1947): A German theoretical physicist. He won the 1918 Nobel Prize in Physics "in recognition of the services he rendered to the advancement of Physics by his discovery of energy quanta." Planck is the originator of the quantum theory.

positron: The antiparticle of the electron. It was proposed by Paul Dirac in 1928 and discovered by Carl D. Anderson in 1932. For this discovery, Anderson won the 1936 Nobel Prize in Physics. In certain experiments with particle accelerators, positrons are used to collide with electrons.

proton: A subatomic particle with positive electric charge. The number of protons in an atom gives its atomic number. It is a hadron and a baryon made up of two up-quarks and one down-quark. The nucleus of hydrogen is one single proton. All other atoms are composed of two or more protons and an equivalent number of neutrons. Protons are stable; the spontaneous decay of free protons has never been observed.

quantum chromodynamics (QCD): The theory of strong interaction between quarks and gluons, which make up hadrons (proton, neutron, or pion). QCD is a quantum field theory and an important part of the Standard Model of physics.

quadrupole magnet: A group of four magnets used in particle accelerators for beam focusing.

quantum: The minimum energy packet that can exist independently; it is a discrete quantity of electromagnetic radiation. This amount of energy is regarded as a unit ("Planck units"). A photon is a single quantum of light and is referred to as a "light quantum."

quantum mechanics: A theory dealing with phenomena on the order of the Planck constant. It provides mathematical descriptions of the particle-like and wavelike behavior of energy and matter. Quantum theory was formulated in the 1920s by Werner Heisenberg, Max Born, Pascual Jordan, Louis de Broglie, Erwin Schrödinger, Wolfgang Pauli, Satyendra Nath Bose, and Niels Bohr.

quantum physics: A branch of physics that deals with discrete, indivisible units of energy called "quanta."

quark: An elementary particle, a fermion, and a constituent of matter. There are six types of quarks. They combine to form hadrons, such as protons and neutrons, the constituents of atomic nuclei. Quarks experience all fundamental forces. The quark model of atoms was proposed by physicists Murray Gell-Mann and George Zweig in 1964.

Rabi, Isidor Isaac: American physicist born in Galicia (1898–1988, 1944 Nobel laureate in Physics "for his resonance method for recording the magnetic properties of atomic nuclei"). He worked at MIT and on the Manhattan Project. Rabi was Science Advisor to President Dwight D. Eisenhower and was involved with the establishment of the Brookhaven National Laboratory (BNL). In 1952, as United States delegate to UNESCO, he became involved with the creation of CERN.

radioactivity (radioactive decay or nuclear decay): The decomposition process of unstable atomic nuclei to form nuclei with a higher stability. There is alpha, beta, and gamma radiation, but there are several other varieties of radioactive decay. The process was discovered by the French scientist Antoine Henri Becquerel (1852–1908) in 1896.

radio-frequency (RF): In accelerator physics, it is a high-frequency alternating voltage that provides or takes energy to or from the particle beam to accelerate or decelerate the particles in the beam.

special relativity: The theory (proposed by Albert Einstein in 1905, in the paper "On the Electrodynamics of Moving Bodies") that the laws of nature are the same for all observers in unaccelerated motion and the speed of light is independent of the motion of its source. The speed of light is a limit; nothing can be faster than the speed of light. Einstein postulated that the time interval between two events was longer for an observer in whose frame of reference the events occur in different places than for the observer for whom they occur at the same place. Special relativity is the most accurate model of motion at any speed.

speed of light: A physical constant denoted by c. Its value is 299,792,458 meters per second (approximately 300,000 kilometers per second) or about 671 million miles per hour. According to Einstein's theory of special relativity, c is the maximum speed of anything traveling in the universe. It may also be the speed of the gravitational waves permeating the universe.

Standard Model: A theory of fundamental particles and how they interact. There are 17 named particles in the Standard Model. The theory concerns the electromagnetic, weak, and strong nuclear interactions. It was developed in the 1960s and 1970s and was finalized after experimental confirmation of the existence of quarks. The theory does not include gravitation. The model has 61 elementary particles. Gauge bosons are defined as force carriers that mediate the strong, weak, and electromagnetic interactions.

strong interaction (force): One of the four fundamental forces or interactions in nature. It is about 100 times stronger than electromagnetism. It is mediated by gluons. The energy of the strong force is used in nuclear power plants and in nuclear weapons as well.

stochastic cooling: A technique in accelerator physics developed by Simon van der Meer at CERN in the 1970s. Stochastic cooling is used to reduce the energy spread of a beam of charged particles. The term "cooling" denotes the reduction of disorder in the beam. Simon van der Meer, together with his colleague Carlo Rubbia, won the 1984 Nobel Prize in Physics "for their decisive contributions to the large project, which led to the discovery of the field particles W and Z, communicators of weak interaction."

synchrotron: A type of circular accelerators. The charged particles are guided by dipole magnets, focused by quadrupole magnets, and accelerated by RF cavities.

synchrotron radiation: Electromagnetic radiation produced when charged particles are accelerated in a curved path by a magnetic field. Synchrotron radiation is also used to study structural details of matter.

Tesla: Unit of magnetic field strength or magnetic flux density (symbol T). The LHC has a peak magnetic field of 8.3 Tesla.

W boson: An elementary particle that—together with the Z boson—mediates the weak interaction. It is named after the weak force. Its discovery at CERN in 1984 was a major step in verifying the Standard Model of physics.

weak interaction: The mechanism responsible for the weak force (interaction). It is one of the four fundamental interactions of nature. The weak interaction is caused by the emission or absorption of W and Z bosons.

weak neutral current: Interaction that is independent of the electric charge of a particle. Particles can exchange energy through this mechanism, but other characteristics of the particles remain unchanged. The weak force, or interaction, is mediated by the Z boson.

WIMPS, "Weakly Interacting Massive Particles": Hypothetical particles serving as one possible solution to the dark matter problem. WIMPS cannot be seen directly and because they do not interact with the strong nuclear force they do not react with "normal" matter.

Z boson: Together with the W boson, the "Z" was discovered in 1983 by physicists at CERN. The Z boson is neutral, and, like the electrically charged W boson, it carries the electroweak force (interaction).

Index